AF389455

Applied Designs in Chemical Structures with High Symmetry

Applied Designs in Chemical Structures with High Symmetry

Special Issue Editors

Lorentz Jäntschi
Beata Szefler

MDPI • Basel • Beijing • Wuhan • Barcelona • Belgrade • Manchester • Tokyo • Cluj • Tianjin

Special Issue Editors
Lorentz Jäntschi
Technical University of
Cluj-Napoca
Romania

Beata Szefler
Nicolaus Copernicus University
Poland

Editorial Office
MDPI
St. Alban-Anlage 66
4052 Basel, Switzerland

This is a reprint of articles from the Special Issue published online in the open access journal *Symmetry* (ISSN 2073-8994) (available at: https://www.mdpi.com/journal/symmetry/special_issues/Applied_Designs_Chemical_Structures_High_Symmetry).

For citation purposes, cite each article independently as indicated on the article page online and as indicated below:

LastName, A.A.; LastName, B.B.; LastName, C.C. Article Title. *Journal Name* **Year**, *Article Number*, Page Range.

ISBN 978-3-03936-571-5 (Hbk)
ISBN 978-3-03936-572-2 (PDF)

Contents

About the Special Issue Editors

Lorentz Jäntschi was born in Fagaras, Romania in 1973. In 1991 he moved to Cluj-Napoca, Cluj, where he completed his studies. In 1995 he was awarded a B.Sc. and M.Sc. in Informatics, in 1997 a B.Sc. and M.Sc. in Physics and Chemistry, in 2000 a Ph.D. in Chemistry under the supervision of Prof. Mircea V. Diudea, in 2002 an M.Sc. in Agriculture, in 2010 a Ph.D. in Horticulture, and, finally, in 2013 a postdoctorate in Horticulture. That same year (2013), he became a Full Profesor of chemistry at the Technical University of Cluj-Napoca and an associate at Babes-Bolyai University, where he advises Ph.D. studies in chemistry. Both positions are to date. During his research and education in Cluj, he performed activities under the auspicies of other institutions as well: G. Baritiu (1995–1999) and Balcescu (1999–2001) at National Colleges, the I. Hatieganu University of Medicine and Pharmacy (2007–2012), Oradea University (2013–2015), and the institute of Agricultural Sciences and Veterinary Medicine at University of Cluj-Napoca (2011–2016). He serves as an editor for the journals Notulae Scientia Biologicae, Notulae Horti Agro Botanici Cluj-Napoca, Open Agriculture and Symmetry. He was Editor-in-Chief of the Leonardo Journal of Sciences and the Leonardo Electronic Journal of Practices and Technologies.

Beata Szefler as a Doctor of Chemistry, is a Professor hab. of Collegium Medicum in Bydgoszcz, Poland, a laboratory diagnostician, and a researcher in the Department of Physical Chemistry, Collegium Medicum, Nicolaus Copernicus University in Bydgoszcz, Poland. A graduate of the Medical Academy in Bydgoszcz, Faculty of Pharmacy, with a Ph.D. in Chemical Sciences from the University of Technology & Life Sciences in Bydgoszcz, Poland, 2010, and awarded habilitation in pharmaceutical sciences from Collegium Medicum, Nicolaus Copernicus University, Bydgoszcz, Poland, in 2019, her experience lies in microbiology, hematology, pharmacokinetics and in silico research (ab initio calculations, HF, DFT, MD, and fullerenes). Her research interests are molecular modeling of aromatic systems with particular emphasis on the importance of biochemical molecules, the study of properties of fullerenes using HF, DFT and MD methods, docking ligand proteins, and QSAR. She is a principal investigator of parameterization of force fields, molecular dynamic simulations of selected protein and ligand–protein systems, quantum-chemical studies of small ligands, the topological description of the electronic structure of ligands (the atoms-in-molecules methodology), and data analysis. She is a member of the European Society of Mathematical Chemistry (ESMC) and the National Chamber of Laboratory Diagnosticians (KIDL, Poland).

Preface to "Applied Designs in Chemical Structures with High Symmetry"

An important factor in determining chemical and biological activity is structural symmetry or asymmetry. Structural symmetry is one of the most basic properties of chemical compounds. The best example of a highly symmetrical compound is C60 fullerene, which appears as a pure synthetic form of carbon. This structure is the best starting compound for the formation of fullerene derivatives with high symmetry. The application of new fullerene derivatives is endless, including in materials science, chemistry, biology, pharmacy and medicine. Thousands of papers on this subject are published annually. Editors Professor Dr. Lorenz Jäntschi and Professor Dr. Beata Szefler are the authors of several dozen scientific papers in which structural symmetry plays a dominant role. The book "Applied Designs in Chemical Structures with High Symmetry" will be a collection of new works describing new structures with these properties.

Lorentz Jäntschi, Beata Szefler
Special Issue Editors

Communication

The Eigenproblem Translated for Alignment of Molecules

Lorentz Jäntschi [1,2]

1   Department of Physics and Chemistry, Technical University of Cluj-Napoca, 400641 Cluj, Romania;
    lorentz.jantschi@chem.utcluj.ro or lorentz.jantschi@ubbcluj.ro or lorentz.jantschi@gmail.com
2   Chemistry Doctoral School, Babeş-Bolyai University, 400084 Cluj-Napoca, Romania

Received: 12 July 2019; Accepted: 7 August 2019; Published: 9 August 2019

Abstract: Molecular conformation as a subproblem of the geometrical shaping of the molecules is essential for the expression of biological activity. It is well known that from the series of all possible sugars, those that are most naturally occurring and usable by living organisms as a source of energy—because they can be phosphorylated by hexokinase, the first enzyme in the glycolysis pathway—are D-sugars (from the Latin dextro). Furthermore, the most naturally occurring amino acids in living cells are L-sugars (from the Latin laevo). However, a problem arises in dealing with the comparison of their conformers. One alternative way to compare sugars is via their molecular alignment. Here, a solution to the eigenproblem of molecular alignment is communicated. The Cartesian system is rotated, and eventually translated and reflected until the molecule arrives in a position characterized by the highest absolute values of the eigenvalues observed on the Cartesian coordinates. The rotation alone can provide eight alternate positions relative to the reflexes of each coordinate.

Keywords: eigenproblem; eigenvalues; molecular alignment; orthogonal alignment

1. Introduction

The topological description of a molecule requires knowledge of the adjacencies (the bonds) between the atoms as well as their identities (the atoms). If this problem is simplified to the extreme, by disregarding the bond types and atom identities, then the adjacencies are simply expressed as 0 or 1 in the vertex adjacency matrix ($[Ad]$) and the identities are expressed as 0 or 1 in the identity matrix ($[Id]$). The characteristic polynomial (ChP) is the natural construction of a polynomial in which the eigenvalues of the $[Ad]$ are the roots of the ChP, as follows:

λ is an eigenvalue of $[Ad] \leftrightarrow$ it follows that $[v] \neq 0$ eigenvector such that $\lambda \cdot [v] = [Ad] \cdot [v]$

$(\lambda \cdot [Id] - [Ad]) \cdot [v] = 0)$; since $v \neq 0 \rightarrow [\lambda \cdot Id - Ad]$ is singular $\rightarrow \det([\lambda \cdot Id - Ad]) = 0$.

Therefore, the characteristic polynomial is defined by:

$$ChP \stackrel{\text{def}}{=} \left| \lambda \cdot [Id] - [Ad] \right|.$$

The characteristic polynomial is a polynomial in λ of the degree of the number of atoms. The eigenproblem (the determination of eigenvalues and eigenvectors) is applicable to any Hessian [1] matrix $[A]$ ($[Ad] \rightarrow [A]$). The mixed derivatives of a scalar-valued function f are the entries off the main diagonal in the Hessian. Assuming that the derivatives are continuous, the order of differentiation does not matter (a result known as Schwarz's, Clairaut's, or Young's theorem), and then the Hessian of f is a symmetric matrix.

Indeed, this is the case (a symmetric matrix) for the (vertex) adjacency matrix, and for the distance matrix—both topological (by bonds) and geometrical (by the atom coordinates).

Related to this problem is the issue of determining the best rotation to relate two sets of vectors. To this issue, a solution was proposed by calculating a symmetric matrix of Lagrange multipliers which is used to minimize the residuals of the linear association between the vectors [2]. Later, different approaches were proposed, such as geometric hashing [3], clique detection [4], the embedding problem [5], Gaussian molecular representation, Gaussian overlap optimization [6], and others covered in [7]. Some of the proposed solutions go a different way, involving physical means forcing the alignment [8,9], while the formulation of similarity metrics was one of the most recently proposed computational alternatives [10]. The alignment serves as a tool for other studies, including similarity analysis [11], docking [12], and structure–activity relationships [13].

The eigenproblem in relation to geometrical alignment was stated before in the context of surface analysis and control [14], and also can go another direction into the context of the molecule. In this context, the molecule is seen as more than a simple unweighted undirected molecular graph with undistinguishable atoms [15].

The eigenproblem of molecular alignment is analyzed in this paper.

2. Materials and Methods

The alignment of molecules can be stated in many ways, as listed in the introduction. For instance, one approach is to search for topological alignment, and another is to search for geometrical alignment. To anticipate the type of molecular alignment, it is necessary to employ the latter method—to search for geometrical alignment.

A molecule is taken here as an example from PubChem CID 444173 ((2R,3S,4R,5R)-oxane-2,3,4,5-tetrol), as shown in Figure 1.

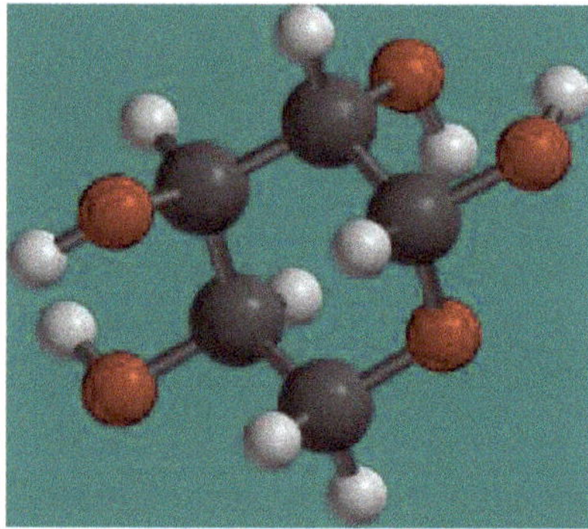

Figure 1. 3D representation of the model of PubChem CID 444173.

For convenience, hydrogen atoms are excluded from the data and the analysis. The next table (Table 1) contains the relevant information for the heavy atoms in the reference molecule.

Table 1. 3D structural data for CID 444173 (heavy atoms, geometric coordinates, and atom symbols).

x (Å)	y (Å)	z (Å)	Atom and Label	
0.7428	−1.4498	−0.0709	O	1
−1.1425	1.1688	1.3882	O	2
1.1461	1.0581	−1.4377	O	3
−2.754	−0.3648	−0.3408	O	4
2.7344	−0.2934	0.3835	O	5
−0.7774	1.0064	0.0187	C	6
0.7504	0.9905	−0.0675	C	7
−1.3475	−0.306	−0.532	C	8
1.3187	−0.2968	0.5474	C	9
−0.671	−1.513	0.1111	C	10

The general way of constructing a characteristic polynomial is to provide an identity matrix [*Id*] and a Hessian matrix (herein labeled as [*A*]). If considering the topology of the molecule, then it is necessary to have the information regarding the connections between the atoms (e.g., bonds). Since all bonds are single bonds for the selected molecule, listing the atoms pairs of the bonds is enough (Table 2).

Table 2. Topology data for CID 444173 (list of bonds between heavy atoms).

(1, 9)	(1, 10)	(2, 6)	(3, 7)	(4, 8)	(5, 9)	(6, 7)	(6, 8)	(7, 9)	(8, 10)

A deeper look into the eigenproblem ($|\lambda \cdot I - A| = 0$) is performed in the next section, with a specific focus on changing of the mathematical properties of the eigenproblem when the adjacencies in [*A*] change from symmetric to anti-symmetric.

3. Results and Discussion

From Table 2, the adjacency matrix [*Ad*] is immediate—zeros represent the entries without a bond between the labeled atoms, while ones appear otherwise. The adjacency matrix is Hessian. For convenience, its characteristic polynomial is:

$$ChP(\lambda; \text{CID444173}, \text{"}[Ad]\text{"}) = 1 \cdot \lambda^{10} - 10 \cdot \lambda^8 + 31 \cdot \lambda^6 - 35 \cdot \lambda^4 + 11 \cdot \lambda^2 - 1 \cdot \lambda^0.$$

As can be seen, the degree of the polynomial is 10, which is equal to the number of the (connected) atoms in the molecule. The general rule is that a characteristic polynomial is always of a degree equal to the size of the square matrices [*I*] and [*A*] (see the before given equation), from which it was derived.

The same strategy can be applied if the adjacency matrix [*Ad*] is replaced by the distance matrix [*Di*]. For convenience, Table 3 lists these two matrices.

Table 3. Adjacency and distance matrices for CID 444173 (heavy atoms).

Ad	1	2	3	4	5	6	7	8	9	10
1	0	0	0	0	0	0	0	0	1	1
2	0	0	0	0	0	1	0	0	0	0
3	0	0	0	0	0	0	1	0	0	0
4	0	0	0	0	0	0	0	1	0	0
5	0	0	0	0	0	0	0	0	1	0
6	0	1	0	0	0	0	1	1	0	0
7	0	0	1	0	0	1	0	0	1	0
8	0	0	0	1	0	1	0	0	0	1
9	1	0	0	0	1	0	1	0	0	0
10	1	0	0	0	0	0	0	1	0	0

Di	1	2	3	4	5	6	7	8	9	10
1	0	4	3	3	2	3	2	2	1	1
2	4	0	3	3	4	1	2	2	3	3
3	3	3	0	4	3	2	1	3	2	4
4	3	3	4	0	5	2	3	1	4	2
5	2	4	3	5	0	3	2	4	1	3
6	3	1	2	2	3	0	1	1	2	2
7	2	2	1	3	2	1	0	2	1	3
8	2	2	3	1	4	1	2	0	3	1
9	1	3	2	4	1	2	1	3	0	2
10	1	3	4	2	3	2	3	1	2	0

It can be checked (but is also true for the general case) that the distance matrix is Hessian, and therefore a characteristic polynomial can be computed for it as well. The next equation lists the *ChP* computed for the distance matrix [*Di*]:

$$ChP(\lambda; \text{CID}444173, "[Di]") = 1 \cdot \lambda^{10} - 313 \cdot \lambda^8 + 3488 \cdot \lambda^7 - 15456 \cdot \lambda^6 - 34720 \cdot \lambda^5 - 40832 \cdot \lambda^4 - 23808 \cdot \lambda^3 - 5376 \cdot \lambda^2$$

The natural extension of this matrix is to employ 3D distances instead of topological distances. Of course, one consequence of this is that the characteristic polynomial would no longer have integer coefficients.

The next table lists the 3D distance matrix (distances were cut to four significant digits) and the roots of the associated characteristic polynomial.

One interesting remark to the data listed in Table 4 is that all roots are real (this is the general behavior for the roots of a characteristic polynomial).

Table 4. *3D* distance matrix and its eigenvalues for CID 444173 (heavy atoms).

3D	3D Distances										Eigenvalues
	1	2	3	4	5	6	7	8	9	10	
1	0	3.541	2.885	3.671	2.347	2.890	2.440	2.427	1.429	1.427	−8.429
2	3.541	0	3.638	2.817	4.264	1.427	2.395	2.430	2.985	3.008	−6.218
3	2.885	3.638	0	4.294	2.769	2.413	1.428	2.983	2.410	3.509	−2.922
4	3.671	2.817	4.294	0	5.536	2.432	3.767	1.421	4.169	2.421	−1.893
5	2.347	4.264	2.769	5.536	0	3.762	2.406	4.183	1.425	3.627	−1.275
6	2.890	1.427	2.413	2.432	3.762	0	1.530	1.533	2.524	2.523	−1
7	2.440	2.395	1.428	3.767	2.406	1.530	0	2.510	1.536	2.884	−0.65
8	2.427	2.430	2.983	1.421	4.183	1.533	2.510	0	2.876	1.526	0
9	1.429	2.985	2.410	4.169	1.425	2.524	1.536	2.876	0	2.372	3.60×10^{-15}
10	1.427	3.008	3.509	2.421	3.627	2.523	2.884	1.526	2.372	0	22.386

Up until this point, the ideas presented in this paper have been reported before. Herein follows the extension to the extant knowledge. What if the same formula is applied to define the *ChP* for Cartesian coordinate distance matrices instead of for the Euclidian distance matrix?

Next three tables (Tables 5–7) list those results (the number of digits is displayed according to the input data—see Table 1).

Table 5. First Cartesian coordinate ("*x*") distances matrix for CID 444173 (heavy atoms).

Dx	1	2	3	4	5	6	7	8	9	10
1	0	1.8853	−0.4033	3.4968	−1.9916	1.5202	−0.0076	2.0903	−0.5759	1.4138
2	−1.8853	0	−2.2886	1.6115	−3.8769	−0.3651	−1.8929	0.2050	−2.4612	−0.4715
3	0.4033	2.2886	0	3.9001	−1.5883	1.9235	0.3957	2.4936	−0.1726	1.8171
4	−3.4968	−1.6115	−3.9001	0	−5.4884	−1.9766	−3.5044	−1.4065	−4.0727	−2.0830
5	1.9916	3.8769	1.5883	5.4884	0	3.5118	1.9840	4.0819	1.4157	3.4054
6	−1.5202	0.3651	−1.9235	1.9766	−3.5118	0	−1.5278	0.5701	−2.0961	−0.1064
7	0.0076	1.8929	−0.3957	3.5044	−1.9840	1.5278	0	2.0979	−0.5683	1.4214
8	−2.0903	−0.2050	−2.4936	1.4065	−4.0819	−0.5701	−2.0979	0	−2.6662	−0.6765
9	0.5759	2.4612	0.1726	4.0727	−1.4157	2.0961	0.5683	2.6662	0	1.9897
10	−1.4138	0.4715	−1.8171	2.0830	−3.4054	0.1064	−1.4214	0.6765	−1.9897	0

Table 6. Second Cartesian coordinate ("y") distances matrix for CID 444173 (heavy atoms).

Dy	1	2	3	4	5	6	7	8	9	10
1	0	−2.6186	−2.5079	−1.0850	−1.1564	−2.4562	−2.4403	−1.1438	−1.1530	0.0632
2	2.6186	0	0.1107	1.5336	1.4622	0.1624	0.1783	1.4748	1.4656	2.6818
3	2.5079	−0.1107	0	1.4229	1.3515	0.0517	0.0676	1.3641	1.3549	2.5711
4	1.0850	−1.5336	−1.4229	0	−0.0714	−1.3712	−1.3553	−0.0588	−0.0680	1.1482
5	1.1564	−1.4622	−1.3515	0.0714	0	−1.2998	−1.2839	0.0126	0.0034	1.2196
6	2.4562	−0.1624	−0.0517	1.3712	1.2998	0	0.0159	1.3124	1.3032	2.5194
7	2.4403	−0.1783	−0.0676	1.3553	1.2839	−0.0159	0	1.2965	1.2873	2.5035
8	1.1438	−1.4748	−1.3641	0.0588	−0.0126	−1.3124	−1.2965	0	−0.0092	1.2070
9	1.1530	−1.4656	−1.3549	0.0680	−0.0034	−1.3032	−1.2873	0.0092	0	1.2162
10	−0.0632	−2.6818	−2.5711	−1.1482	−1.2196	−2.5194	−2.5035	−1.2070	−1.2162	0

Table 7. Third Cartesian coordinate ("z") distances matrix for CID 444173 (heavy atoms).

Dz	1	2	3	4	5	6	7	8	9	10
1	0	−1.4591	1.3668	0.2699	−0.4544	−0.0896	−0.0034	0.4611	−0.6183	−0.1820
2	1.4591	0	2.8259	1.7290	1.0047	1.3695	1.4557	1.9202	0.8408	1.2771
3	−1.3668	−2.8259	0	−1.0969	−1.8212	−1.4564	−1.3702	−0.9057	−1.9851	−1.5488
4	−0.2699	−1.7290	1.0969	0	−0.7243	−0.3595	−0.2733	0.1912	−0.8882	−0.4519
5	0.4544	−1.0047	1.8212	0.7243	0	0.3648	0.4510	0.9155	−0.1639	0.2724
6	0.0896	−1.3695	1.4564	0.3595	−0.3648	0	0.0862	0.5507	−0.5287	−0.0924
7	0.0034	−1.4557	1.3702	0.2733	−0.4510	−0.0862	0	0.4645	−0.6149	−0.1786
8	−0.4611	−1.9202	0.9057	−0.1912	−0.9155	−0.5507	−0.4645	0	−1.0794	−0.6431
9	0.6183	−0.8408	1.9851	0.8882	0.1639	0.5287	0.6149	1.0794	0	0.4363
10	0.1820	−1.2771	1.5488	0.4519	−0.2724	0.0924	0.1786	0.6431	−0.4363	0

It can be observed that the Cartesian coordinates distance matrices are no longer symmetric matrices, but are in fact anti-symmetric, meaning that $M_{i,j} = -M_{j,i}$.

The beauty of the result shown by taking a look at the eigenvalues. The next table (Table 8) lists the eigenvalues for all matrices.

Table 8. Eigenvalues for CID 444173 (heavy atoms).

	x_1	x_2	x_3	x_4	x_5	x_6	x_7	x_8	x_9	x_{10}
[Ad]	−2.318	−1.556	−1.334	−0.506	−0.41	0.41	0.506	1.334	1.556	2.318
[Di]	−8.429	−6.218	−2.922	−1.893	−1.275	−1	−0.65	0	3.60×10^{-15}	22.386
[3D]	−8.722	−5.093	−3.145	−2.521	−1.474	−1.257	−1.05	−0.911	−0.784	25.007
[Dx]	15.299·i	−15.299·i	0	0	0	0	0	0	0	0
[Dy]	9.629·i	−9.629·i	0	0	0	0	0	0	0	0
[Dz]	6.973·i	−6.973·i	0	0	0	0	0	0	0	0

It should be noted that the values listed in Table 7 reveal some computational errors. It is obvious (and it is so) that 3.6×10^{-15} is actually a "0" and it is necessary to be aware of this type of error coming from "machine epsilon" [16] which is about 10^{-7} for "single" precision, 10^{-16} for "double" precision, and about 10^{-19} for "extended" precision. Most floating-point implementations use "double" precision and thus the listed value (3×10^{-15}) "fits in range".

More important, as can be observed (see Table 7), the eigenvalues of [Dx], [Dy], [Dz] are all 0 excepting (always) two—which are (always) paired and (always) imaginary (i = $\sqrt{-1}$, see Table 8). This is the opposite of the traditional case of symmetric matrices, when the values are (always) real. This is the beauty of the result.

Moreover, it should be noted that the polynomial can be expressed with real-value coefficients as a product of a polynomial of degree 2 and a monomial of degree ($n − 2$), as listed in Table 9.

Table 9. The polynomials of $[Dx]$, $[Dy]$, and $[Dz]$ for CID 444173 (heavy atoms).

Matrix (A)	$\|\lambda \cdot I - A\|$ Polynomial
$[Dx]$	$\lambda^8 \cdot (\lambda^2 + 234.0448052)$
$[Dy]$	$\lambda^8 \cdot (\lambda^2 + 92.7157814)$
$[Dz]$	$\lambda^8 \cdot (\lambda^2 + 48.6224414)$

A consequence is hidden behind this result—to obtain those two coefficients (which are actually the first and third coefficients, independent of how many atoms are in the molecule) it is necessary to obtain their roots. Therefore, it is not necessary to run an "eigenvalues" routine to obtain them; it is enough to run only two steps of a coefficient determination program (such as that described in [17]), which will produce a result much more quickly.

So, what if we conduct a rotation of the molecule?

For example, by rotating the molecule by 15° (15/180 radians; coordinates are given in Table 1), the values for the polynomials are changed—see Table 10. First it should be pointed out that the polynomial is no longer invariant due to the choice of the system of coordinates. If invariants are sought, this is not a good situation—but for the purpose of addressing the alignment problem, this setup is very useable.

Table 10. The polynomials of $[Dx]$, $[Dy]$, and $[Dz]$ rotated (15°, 15°, 15°) for CID 444173 (heavy atoms).

Matrix (A)	$\|\lambda \cdot I - A\|$ Polynomial
$[Dx]$	$\lambda^8 \cdot (\lambda^2 + 162.836846)$
$[Dy]$	$\lambda^8 \cdot (\lambda^2 + 90.150945)$
$[Dz]$	$\lambda^8 \cdot (\lambda^2 + 47.28921)$

In the general case, with a_1, a_2, and a_3 as rotation angles defining the rotation matrices (given below), it is necessary to maximize the variance along the axes of coordinates.

$$\begin{bmatrix} 1 & 0 & 0 \\ 0 & \cos(a_2) & \sin(a_2) \\ 0 & -\sin(a_2) & \cos(a_2) \end{bmatrix}, \begin{bmatrix} \cos(a_1) & 0 & -\sin(a_1) \\ 0 & 1 & 0 \\ \sin(a_1) & 0 & \cos(a_1) \end{bmatrix}, \begin{bmatrix} \cos(a_0) & \sin(a_0) & 0 \\ -\sin(a_0) & \cos(a_0) & 0 \\ 0 & 0 & 1 \end{bmatrix}$$

This results a two-step algorithm, described below:

- Since rotation by a_0 leaves untouched the "z" coordinate, the first problem is to find a value of a_0 such that the squared sum of the eigenvalue(s) for the $[Dx]$ matrix is minimized (or its coefficient from Table 9, which is $x_1 \cdot x_2 = x_1 \cdot \overline{x}_1 = -x_1{}^2 = -x_2{}^2$, is maximized);
- Next, we need to leave untouched the "x" coordinate—which was already fitted in the first step. For this, we may want to employ rotation by a_2, such that the squared sum of the eigenvalue(s) for the $[Dy]$ matrix is minimized (or its coefficient from Table 9 is maximized);
- There is no third step involving the third rotation matrix, because by maximizing (or minimizing) the first two coordinates, we have already employed all coordinates (x and y in the first step; y and z in the second).

Therefore, at this point we have the alignment of the molecule.

The problem of molecular 3D alignment involving the modified characteristic polynomial (eigenproblem) becomes a combinatorial problem since, after eigenvector minimization by each (two out of three) Cartesian coordinate, we obtain the molecules in their proper alignment or in the mirror of the proper alignment, when "$x_i \leftarrow -x_i$" and/or "$y_i \leftarrow -y_i$" and/or "$z_i \leftarrow -z_i$" transformation will align it.

Of course, a question may arise: what is the meaning of such alignment? This research is ongoing, but so far it has been found that this alignment corresponds to the minimization of the rotation inertia of the coordinates. In other words, the thinnest part of the molecule aligns with one coordinate, and then the thinnest part of what remains (so the molecule can be rotated around that axis) aligns with the second coordinate.

Revising the results communicated here, it should be noted that the classical eigenproblem is addressed to symmetric matrices—such as are the topological adjacency and topological distance matrices (shown in Table 3) and the geometrical distance matrix (Table 4). The peculiarity of the Cartesian distance matrices (shown in Tables 5–7) is the fact that they are anti-symmetric, sometimes called skew-symmetric matrices. This is, in mathematical terms, a strong property—as strong as the property of symmetry (please note that here the symmetry describes the matrices—namely, matrix A is symmetric if $A = A^{\mathrm{T}}$ and it is anti-symmetric if $A = -A^{\mathrm{T}}$). On the other hand, the elements of the Cartesian coordinate matrices are mirrored relative to the main diagonal—this property is called reflection symmetry, line symmetry, or mirror symmetry—which makes these matrices very suitable for the same set of operations that are typically employed for symmetric matrices. Further, among the known properties of skew-symmetric matrices is the fact illustrated in Table 8—If A is a real skew-symmetric matrix and λ is a real eigenvalue, then $\lambda = 0$, i.e., the nonzero eigenvalues of a skew-symmetric matrix are purely imaginary". Since a skew-symmetric matrix is similar to its own transposition, they must have the same eigenvalues. It follows that the eigenvalues (λ) of a skew-symmetric matrix always come in pairs ($\pm\lambda$), a property which is also illustrated in Table 8.

It should be noted that the generation of Cartesian coordinates from the diagonalization of adjacency or distance-related matrices is quite standard in mathematical chemistry. For instance, the methods to generate fullerene cages from Schlegel diagrams are normally embedded in fullerene sw packages (see for example [18]). Thus, the results communicated here may have useful applications in this regard.

4. Conclusions

The change from symmetry to anti-symmetry in the adjacency matrix of the eigenproblem moves the eigenvalues from real space into imaginary space. When the eigenequation is applied to the Cartesian space of the molecule instead of the topological or Euclidean spaces, the resultant roots (corresponding to the eigenvalues) are all 0 (multiple roots) excepting two, which are always imaginary (and complementary). The rotation of a molecule induces into the Cartesian space a way of aligning the molecule by maximizing the magnitude of the roots in a preselected order of the Cartesian axes. This property can be further exploited for the alignment of multiple molecules, when for highly symmetric molecules the alignment problem is turned into the $(S_2)^3$ conformational problem.

Though the programs provided in the Supplementary Materials can be used to align any molecule, they are not communicated as a novel tool. Aligning a molecule by its Cartesian coordinates via the simultaneous alignment of many molecules—such as for molecular docking purposes—will require further study.

Supplementary Materials: The datafile for CID 444173 and the MATLAB program implementing the Cartesian alignment of a molecule are available online at http://www.mdpi.com/2073-8994/11/8/1027/s1.

Author Contributions: The author designed and made the study and also wrote the paper.

Funding: This research received no external funding.

Conflicts of Interest: The author declares no conflict of interest.

References

1. Sylvester, J.J. On the theorem connected with Newton's rule for the discovery of imaginary roots of equations. *Messenger Math.* **1880**, *9*, 71–84.
2. Kabsch, W. A solution for the best rotation to relate two sets of vectors. *Acta Crystallogr.* **1976**, *A32*, 922–923. [CrossRef]

3. Lamdan, Y.; Wolfson, H.J. Geometric Hashing: A General and Efficient Model-based Recognition Scheme. In Proceedings of the Second International Conference on Computer Vision Proceedings, Tampa, FL, USA, 5–8 December 1988; pp. 238–249.
4. Martin, Y.C.; Bures, M.G.; Danaher, E.A.; DeLazzer, J.; Lico, I.; Pavlik, P.A. A fast new approach to pharmacophore mapping and its application to dopaminergic and benzodiazepine agonists. *J. Comput. Aided Mol. Des.* **1992**, *7*, 83–102. [CrossRef]
5. Mrippen, G.M.; Havel, T.F. *Distance Geometry and Molecular Conformation*; Research Studies Press: Taunton, UK; Wiley: New York, NY, USA, 1988; pp. 1–541.
6. Kearsley, S.K.; Smith, G.M. An alternative method for the alignment of molecular structures: Maximizing electrostatic and steric overlap. *Tetrahedron Comput. Methodol.* **1990**, *3*, 615–633. [CrossRef]
7. Lemmen, C.; Lengauer, T. Computational methods for the structural alignment of molecules. *J. Comput. Aided Mol. Des.* **2000**, *14*, 215–232. [CrossRef] [PubMed]
8. Friedrich, B.; Herschbach, D. Alignment and trapping of molecules in intense laser fields. *Phys. Rev. Lett.* **1995**, *74*, 4623–4626. [CrossRef] [PubMed]
9. Larsen, J.J.; Hald, K.; Bjerre, N.; Stapelfeldt, H.; Seideman, T. Three dimensional alignment of molecules using elliptically polarized laser fields. *Phys. Rev. Lett.* **2000**, *85*, 2470–2473. [CrossRef] [PubMed]
10. Cai, C.; Gong, J.; Liu, X.; Gao, D.; Li, H. SimG: An alignment based method for evaluating the similarity of small molecules and binding sites. *J. Chem. Inf. Model.* **2013**, *53*, 2103–2115. [CrossRef]
11. Wang, N.; DeLisle, R.K.; Diller, D.J. Fast small molecule similarity searching with multiple alignment profiles of molecules represented in one-dimension. *J. Med. Chem.* **2005**, *48*, 6980–6990. [CrossRef]
12. Benyamini, H.; Shulman-Peleg, A.; Wolfson, H.J.; Belgorodsky, B.; Fadeev, L.; Gozin, M. Interaction of C60-fullerene and carboxyfullerene with proteins: Docking and binding site alignment. *Bioconj. Chem.* **2006**, *17*, 378–386. [CrossRef]
13. Cramer, R.D., III; Patterson, D.E.; Bunce, J.D. Comparative molecular field analysis (CoMFA). 1. Effect of shape on binding of steroids to carrier proteins. *J. Am. Chem. Soc.* **1988**, *110*, 5959–5967. [CrossRef]
14. Huang, J.; Zhang, M.; Ma, J.; Liu, X.; Kobbelt, L.; Bao, H. Spectral Quadrangulation with Orientation and Alignment Control. In Proceedings of the ACM SIGGRAPH Asia 2008 Papers, Singapore, 10–13 December 2008.
15. Joiţa, D.M.; Jäntschi, L. Extending the Characteristic Polynomial for Characterization of C20 Fullerene Congeners. *Mathematics* **2017**, *5*, 84. [CrossRef]
16. Goldberg, D. What every computer scientist should know about floating-point arithmetic. *ACM Comput. Surv.* **1991**, *23*, 5–48. [CrossRef]
17. Jäntschi, L.; Bolboacă, S.D. 7. Characteristic Polynomial (CHARACT-POLY). In *New Frontiers in Nanochemistry: Concepts, Theories, and Trends, 3-Volume Set: Volume 2: Topological Nanochemistry*; Putz, M.V., Ed.; Apple Academic Press: Waretown, NJ, USA, 2019; pp. 95–117.
18. Schwerdtfeger, P.; Wirz, L.; Avery, J. Fullerene—A software package for constructing and analyzing structures of regular fullerenes. *J. Comput. Chem.* **2013**, *34*, 1508–1526. [CrossRef]

Article

Docking Linear Ligands to Glucose Oxidase

Beata Szefler

Department of Physical Chemistry, Faculty of Pharmacy, Collegium Medicum, Nicolaus Copernicus University, Kurpińskiego 5, 85-096 Bydgoszcz, Poland; beatas@cm.umk.pl

Received: 12 June 2019; Accepted: 8 July 2019; Published: 10 July 2019

Abstract: GOX (3QVR), glucose oxidase, is an oxidoreductase enzyme, which has found many applications in biotechnology and modern diagnostics with typical assays including biosensors useful in the determination of free glucose in body fluids. PEI (polyethylenimines) are polymer molecules made up of amine groups and two aliphatic carbons, which are cyclically repeated. PEI are transfection reagents which, using positively charged units, bind well to anionic DNA residues. During the studies on GOX, PEI were used both in their linear and branched structures. Rhombellanes, RBL, are structures decorated with rhombs/squares. The aim of the paper is to study the interactions of two kinds of linear ligands: PEIs (Polyethylenimines) and CHRs (ethers of Hexahydroxy-cyclohexane) with the glucose oxidase enzyme, GOX (3QVR). To understand the structure-activity relationship between the GOX enzyme and the linear ligands PEI and CHR, two steps of docking simulation were performed; mapping the whole area of the 3QVR enzyme and docking on the first and second surface of the enzyme, separately. The studied ligands interacted with amino acids of GOX inside the protein and on its surface, with stronger and shorter bonds inside of the protein. However, long chain ligands can only interact with amino acids on the external protein surface. After the study, two domains of the enzyme were clearly evidenced; the external surface domain more easily creates interactions with ligands, particularly with CHR ligands.

Keywords: PEI; CHR; 3QVR; Glucose oxidase; docking

1. Introduction

GOX, glucose oxidase, is an oxidoreductase that catalyzes the oxidation of β-D-glucose to D-glucono-β-lactone, which is non-enzymatically hydrolyzed to gluconic acid [1]. In these reactions, FADH2 is formed as the reduction product of the FAD ring of GOX. FADH2 is then re-oxidized by using molecular oxygen to obtain hydrogen peroxide (H_2O_2). GOX has found many applications in biotechnology and modern diagnostics. Typical assays utilize it as a biosensor useful in determination of free glucose in body fluids.

There are several types of PEI [2]: The branched PEI (BPEI), linear PEI (LPEI), and dendrymer PEI (DPEI). Branched PEI have all types of amino groups, while linear PEI contain primary and secondary amino groups.

BPEI is liquid at room temperature, while LPEI is solid because its melting point is about 73–75 °C. It is well soluble in water (hot with low pH), methanol, ethanol, and chloroform.

PEI binds to anionic residues of DNA by its positively charged units [3]. Despite high toxicity, PEI has many applications [4] due to its polycationic character. PEI has been used in studies on GOX, both in its linear and branched structures [5–8].

Rhombellanes, RBL, are structures built of rings in the form of rhombs or squares (Figure 1, left); they have recently been proposed by Diudea [9,10].

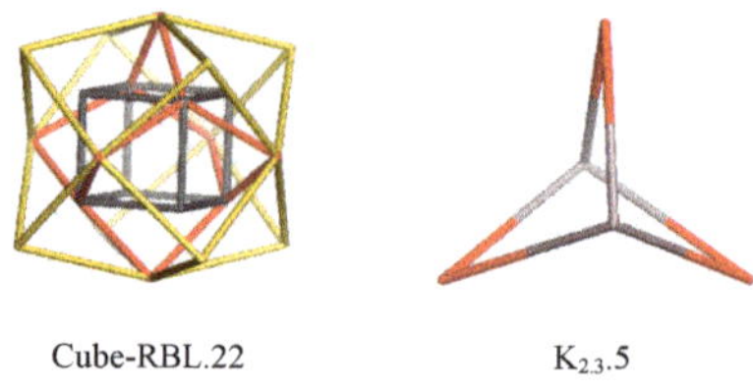

Cube-RBL.22 $K_{2.3}.5$

Figure 1. Rhombellane basic structures [9,10].

Rombellanes are organic molecules, C5H6, structurally related to [1,1,1] propellane, which contains only triangles [11]. Its reduced form, C5H8, is named bicyclo [1.1.1]pentane and is made of rhomb rings or square rings (Figure 1, right). [1,1,1] Propellane undergoes spontaneous polymerization to [n] staffanes (Figure 2, left) [12,13], which are rigid, linear structures. The number suffixing a molecule name stands for the number of its (heavy) atoms.

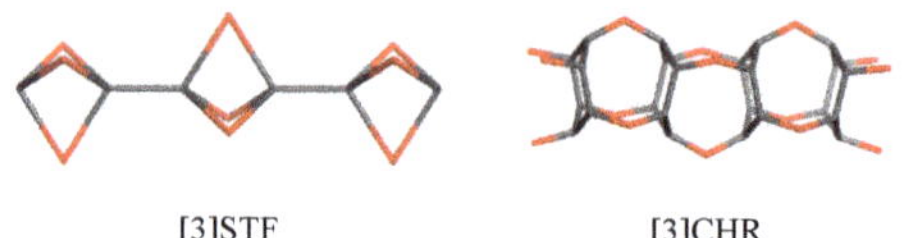

[3]STF   [3]CHR

Figure 2. Linear oligomers related to rhombellanes [9,10].

Rhombellanes have five common features [14–16], but foremost they are structures made of rhomb rings or square rings, where vertex classes consist of all non-connected vertices. The cube-rhombellane is presented in Figure 1. Rhombellanes are, in general, designed by the "rhombellation" operation [9,10].

By analogy to staffanes [n]STF [5], (Figure 2, left), a linear rod-like polymer (a poly-ether of 1,2,3,4,5,6-hexahydroxy-cyclohexane, [n] CHR, (Figure 2, right) was proposed by Diudea [17]. This idea came from cube-rhombellane (Figure 1, right) [9,10], with vertex/atom connectivity 6 and 3, respectively. For its realization as a molecule, it was proposed [10] to use hexahydroxy-cyclohexane for connectivity 6 (connectivity 3 being more accessible).

2. Materials and Methods

In the first step of the docking simulation, the whole area of the 3QVR enzyme was mapped step by step (Figure 3). The protein was downloaded from Brookhaven Protein Database PDB [18,19].

To understand the structure–activity relationship of the ligand–protein complex (Supplementary Tables S1 and S2), molecular docking was carried out.

Two kinds of linear structures were used as ligands: $PEI.C_{2+n}N_{2+m}$; $n = 2, 4, 6 \ldots$; $m = 1, 2, 3 \ldots$ (Figure 4) and [n]CHR.$24 + 15n$; $n = 0, 1, 2 \ldots$, respectively (Figure 4).

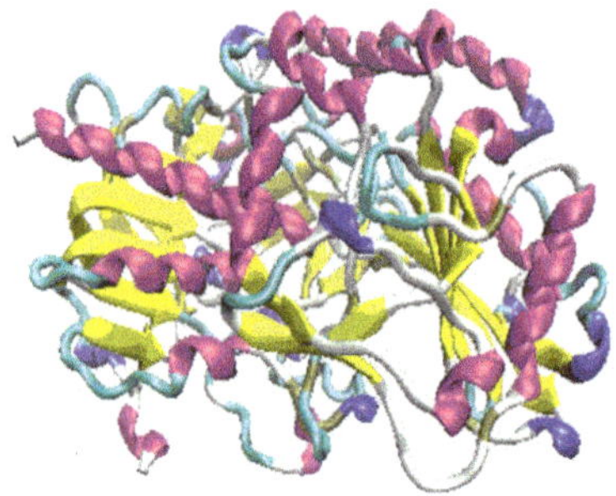

Figure 3. *Cont.*

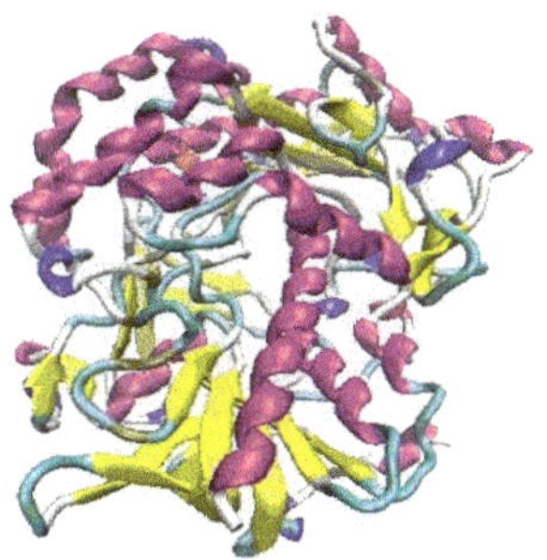

Figure 3. The protein 3QVR/GOX, with the inside pocket (**top**) and external binding domain (**bottom**) docking surface.

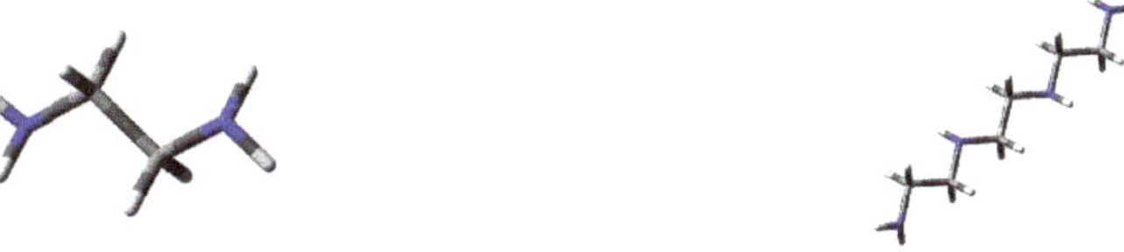

Linear ligand PEI: a unit (left) and an oligomer_$C_{2+n}N_{2+m}$ (right)

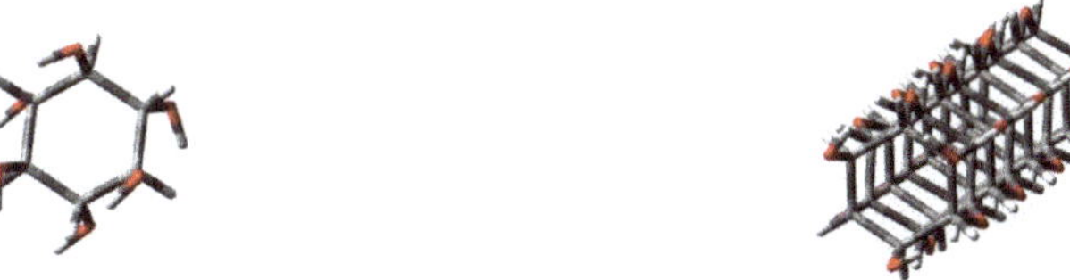

Linear ligand CHR: 1,2,3,4,5,6-Hexahydroxy-cyclohexane, HCH (left) and an oligomer (right).

Figure 4. Linear ligands PEI (**top row**) and CHR (**bottom row**).

All structures used during the molecular docking stage, including the ligands and proteins, were created with the use of the AutoDockTools package and non-polar hydrogen atoms were removed from each molecule. The dimensions of the grid box were fitted to the size of the protein; while during modeling, all possible interactions of the ligand with the whole protein surface were evaluated. The docking procedure was realized with the use of AutoDockVina [20,21], using the Lamarckian genetic algorithm [22]. The calculations were realized with exhaustiveness parameter equal 20, with the chosen value being a compromise between good reproducibility of calculations and computation time cost. Finally, the validity of the used algorithm was verified [23,24].

3. Results

3.1. The First Step of the Study

By performing the docking procedure, it is clear that ligands can interact with amino acids of GOX both inside of protein (Figure 5, top) and on its surface (Figure 5, bottom). Selected values of the free energy of binding to the proteins active site are given in Supplementary Tables S1 and S2 for ligands PEI and CHR, respectively.

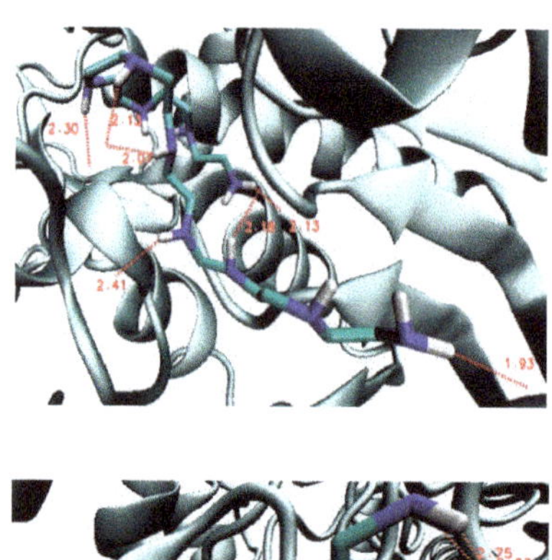

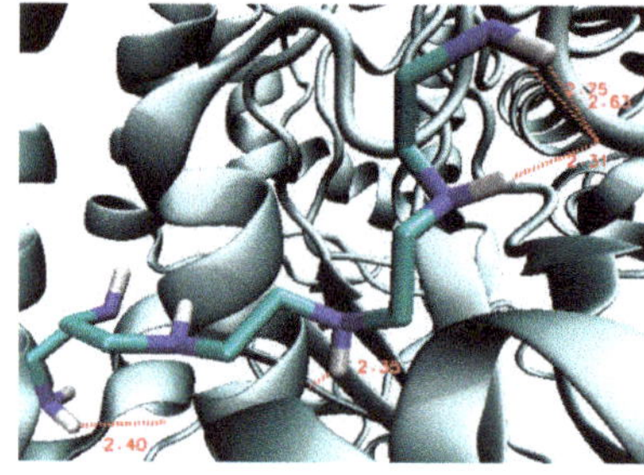

Figure 5. Hydrogen bond lengths (in Å) in the PEI–GOX complexes: Inside the protein (3QVR–PEI.C$_{18}$N$_{10}$.01.0Lin2, **top**) and on its surface (3QVR–PEI.C$_{10}$N$_6$.01.0Lin3, **bottom**).

However, the bonds created inside are stronger and shorter, compared with the bonds formed on the outside surface of enzyme, which is what could be expected (see Supplementary Tables S1 and S2; Supplementary Figures S1 and S2. The structural analysis of the formed complexes clearly shows that the ligand affinity values on the external surface are lower compared to the values of affinity inside of enzyme (see also Supplementary Figures S1–S3; Supplementary Tables S1 and S2.

With the increasing chain length of PEI and CHR ligands, the values of ligand–protein affinity also increase (Supplementary Figure S4; Supplementary Tables S1 and S2. This is caused by the fact that increasing the structural elongation of the ligand increases the number of atoms participating in the ligand–enzyme hydrogen bond formation (Figure 6).

Hydrogen bonds of ligand and protein are detailed in Supplementary Figures S1 and S2 (amino acids are hidden for the sake of image clarity).

The ligand affinity depends on the number of hydrogen bonds and the strength of the interactions between ligand and protein (Figure 7; Supplementary Table S3). It also depends on whether these interactions occur on the external surface of the enzyme or inside it.

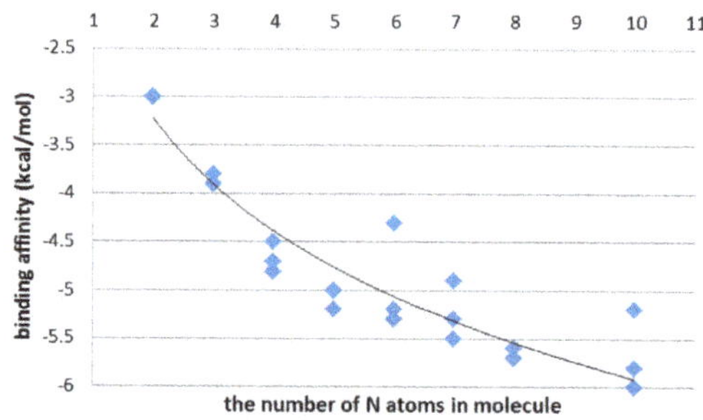

Figure 6. *Cont.*

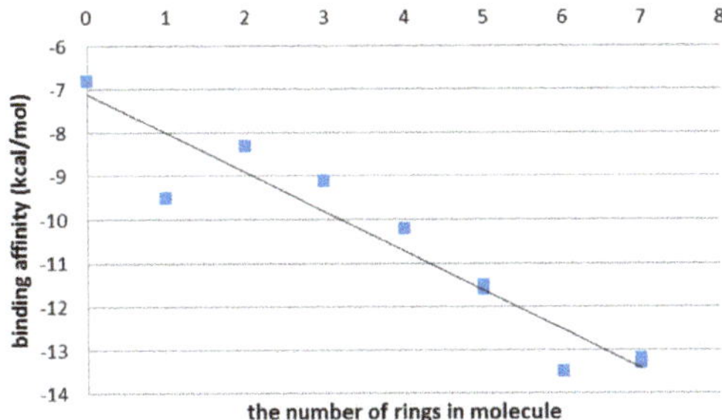

Figure 6. Binding affinity (kcal/mol) in PEI (**top**) and CHR (**bottom**) ligands, after the first step of the docking procedure.

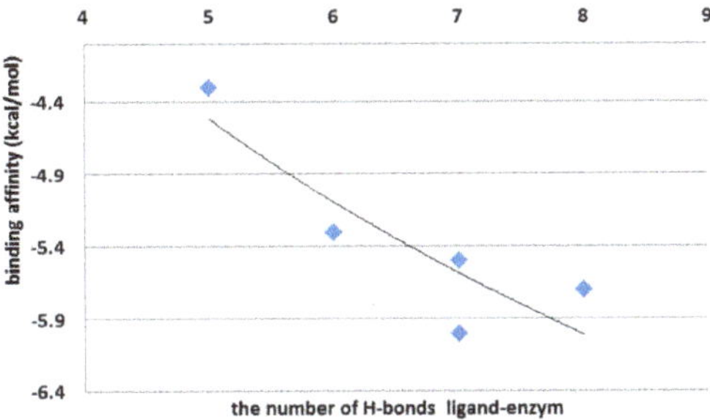

Figure 7. Binding affinity (kcal/mol) vs. the number of H-bonds created between ligand and enzyme.

Jeffrey [25] categorizes H bonds by donor–acceptor distances. Based on such categorization, H-bonds created with the protein by PEI and CHR ligands are strong and moderate (Supplementary Table S3). The number and strength of the hydrogen bonds between the ligand and protein will, finally, determine the values of binding affinity, and consequently, the ligand–protein overall distance. Thus, a high value of ligand–protein binding affinity is directly related to the formation of many strong hydrogen bonds.

In the case of PEI ligands, all the tested molecules, except for $PEI.C_{10}N_6.01.0Lin3$, form H-bonds inside of the protein. In the case of CHR molecules, only those with a short chain (namely [0]CHR.24 and [1]CHR.39) find room to interact with amino acids inside of the enzyme. Ligands with a longer chain ([n]CHR.24 + 15n; n = 3 to 7) can only embed on the external binding domain of the enzyme surface (Figure 3).

That is why, in the second step of this study, the docking was made on the surface of the enzyme, where the studied ligands can create H-bonds with the protein in an easy and quick way.

3.2. The Second Step of the Study

Again, with the growing chain length of PEI and CHR molecules, the values of ligand–protein affinity also increase (Figure 8), by increasing the number of hydrogen bonds created between the enzyme and ligand.

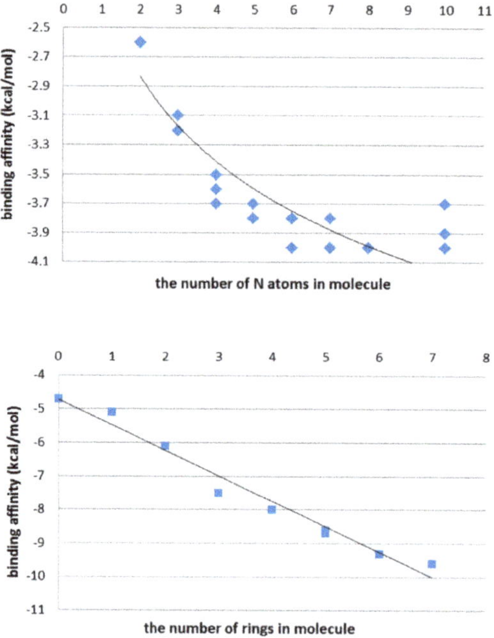

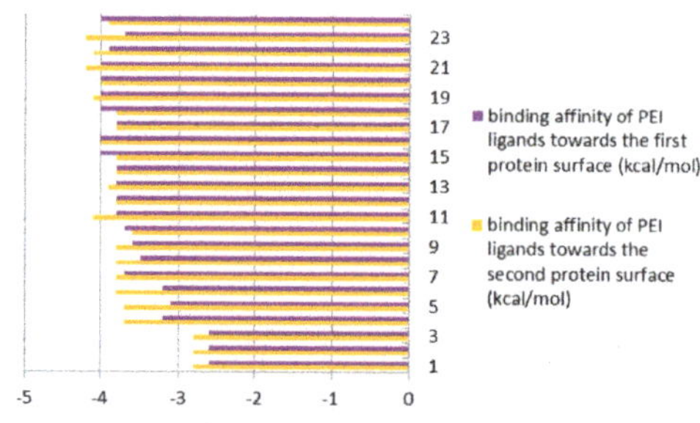

Figure 8. Binding affinity (kcal/mol) in PEI (**top**) and CHR molecules (**bottom**) after docking of ligands on the external surface of protein.

One can see that the values of ligand–enzyme affinity on the external domain (Supplementary Tables S4 and S5) are much smaller than in the inside domain (Supplementary Tables S6 and S7), which suggests the creation of longer and weaker hydrogen bonds.

4. Discussion

When comparing the energy of hydrogen bonds created between ligands and GOX enzyme, it can be found that differences were around 1% in favor of the external domain of docking (Figures 4 and 9, top) for PEI ligands; while in the case of CHR ligands, the differences ranged from 40% to 70%, also in favor of the external domain (Figure 9, bottom).

All of the above means that the external surface domain of enzyme (Figure 3, bottom) interacts more easily with ligands, particularly with CHR molecules. Here, the larger differences are in the case of longer chain molecules (Figure 9, bottom).

Figure 9. *Cont.*

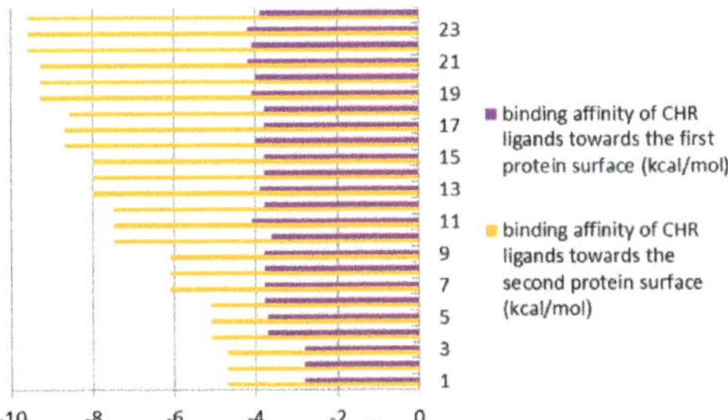

Figure 9. Binding affinities (in kcal/mol) of enzyme (GOX protein)–ligands (PEI (**top**) and CHR (**bottom**)), bound inside (first protein surface) and outside/externally (second protein surface), respectively.

Detailed structural analysis of the ligand-protein complexes, formed after the docking procedure, confirmed the observation regarding the existing differences in the strength of hydrogen bonds, on the inside and outside domains of GOX enzyme, in docking of the ligands PEI and CHR, respectively (Figure 9). The number and strength of hydrogen bonds created between PEI and amino acids were comparable on the two docking domains of the protein. Contrarily, in the case of CHR ligands, the external surface domain provided more room to form a large number of interactions with the protein, in comparison to the internal pocket of enzyme (Figure 3).

5. Conclusions

To understand the structure–activity relationship between the GOX (3QVR) enzyme and linear ligands PEI and CHR, in the first step of this docking study, the whole area of the enzyme was mapped. It was found that ligands can interact with amino acids of GOX in two domains, namely, the inside of the protein and on its external surface. The bonds created in the inside domain are stronger and shorter, compared with the bonds formed on the external surface of enzyme. Also, in the latter, the bonding energy values were lower compared with the values recorded for the complexes formed with the inside docking domain of enzyme. Despite this, with the increasing chain length of PEI and CHR ligands, the values of ligand–protein affinity also increased, which is what is connected to the increasing ability to create more ligand–protein hydrogen bonds. The binding affinity depends on the number and strength of interactions between ligand and protein, and also depends on whether these interactions occur on the external surface of the enzyme or inside it. The longer ligand chains can only find room for docking with the amino acids lying on the external surface of the protein.

Comparison between the two docking sites of GOX revealed differences in binding affinity in favor of the external domain of docking; around 1% in the case of PEI ligands and from 40% to 70% in the case of CHR molecules.

The number and strength of hydrogen bonds created between PEI and GOX amino acids are comparable on the two sites of protein docking; while in case of CHR ligands, the docking is favored on the external domain of the protein surface. This means that the CHR ligands act selectively on the external binding site of the GOX enzyme.

Supplementary Materials: The following are available online at http://www.mdpi.com/2073-8994/11/7/901/s1, Figure S1: The length of hydrogen bonds in Å formed between ligands PEI and GOX enzyme. Table S1: The final lamarckian genetic algorithm docked state—Binding energy of PEI ligand to the active sites of type of 3QVR during the nine explored conformations. The number 1,2,3 in the ligand name means the use of another random variable in the docking procedure.

Funding: This research was supported by PL-Grid Infrastructure http://www.plgrid.pl/en.

Acknowledgments: This work was supported by GEMNS project granted in the European Union's Seventh Framework Programme under the frame of the ERA-NET EuroNanoMed II (European Innovative Research and Technological Development Projects in Nanomedicine).

Conflicts of Interest: The author declares no conflict of interest.

References

1. Leskovac, V.; Trivić, S.; Wohlfahrt, G.; Kandrac, J.; Pericin, D. Glucose oxidase from Aspergillus niger: The mechanism of action with molecular oxygen, quinines and one electron acceptors. *Int. J. Biochem. Cell Biol.* **2005**, *37*, 731–750. [CrossRef] [PubMed]
2. Yemul, O.; Imae, T. Synthesis and characterization of poly(ethyleneimine) dendrimers. *Colloid Polym. Sci.* **2008**, *286*, 747–752. [CrossRef]
3. Rudolph, C.; Lausier, J.; Naundorf, S.; Müller, R.H.; Rosenecker, J. In vivo gene delivery to the lung using polyethylenimine and fractured polyamidoamine dendrimers. *J. Gene Med.* **2000**, *2*, 269–278. [CrossRef]
4. Akinc, A.; Thomas, M.; Klibanov, A.M.; Langer, R. Exploring polyethylenimine-mediated DNA transfection and the proton sponge hypothesis. *J. Gene Med.* **2004**, *7*, 657–666. [CrossRef] [PubMed]
5. Szefler, B.; Diudea, M.V.; Grudziński, I.P. Nature of polyethylene imine-glucose oxidase interactions. *Studia UBB_Chemia* **2016**, *61*, 249–260.
6. Lungu, C.N.; Diudea, M.V.; Putz, M.V.; Grudziński, I.P. Linear and branched PEIs (Polyethylenimines) and their properties space. *Int. J. Mol. Sci.* **2016**, *17*, 555. [CrossRef] [PubMed]
7. Szefler, B.; Diudea, M.V.; Putz, M.V.; Grudziński, I.P. Molecular Dynamic Studies of the Complex Polyethylenimine and Glucose Oxidase. *Int. J. Mol. Sci.* **2016**, *17*, 1796. [CrossRef]
8. Lungu, C.N.; Diudea, M.V.; Putz, M.V.; Grudzinski, I.P. FAD molecular adaptability among surrounding amino acids and its catalytic role in glucose oxidase and related flavoproteins. *Res. J. Life Sci. Bioinf. Pharm. Chem. Sci.* **2017**, *3*. [CrossRef]
9. Diudea, M.V. Rhombellanes—A new class of structures. *Int. J. Chem. Model.* **2017**, *9*, 91–96.
10. Diudea, M.V. Cube-Rhombellane: From graph to molecule. *Int. J. Chem. Model.* **2017**, *9*, 97–103.
11. Wiberg, K.B.; Walker, F.H. [1.1.1] Propellane. *J. Am. Chem. Soc.* **1982**, *104*, 5239–5240. [CrossRef]
12. Kazynsky, P.; Michl, J. [n] Staffanes: A molecular-size tinkertoy construction set for nanotechnology. Preparation of end-functionalized telomers and a polymer of [1.1.1] propellane. *J. Am. Chem. Soc.* **1988**, *110*, 5225–5226. [CrossRef]
13. Dilmaç, A.; Spuling, E.; Meijere, A.; Bräse, S. Propellanes—From a chemical curiosity to "explosive" materials and natural products. *Angew. Chem. Int. Ed.* **2017**, *56*, 5684–5718. [CrossRef] [PubMed]
14. Diudea, M.V. Hypercube related polytopes. *Iran. J. Math. Chem.* **2018**, *9*, 1–8.
15. Diudea, M.V. Rhombellanic crystals and quasicrystals. *Iran. J. Math. Chem.* **2018**, *9*, 167–178.
16. Szefler, B.; Czeleń, P.; Diudea, M.V. Docking of indolizine derivatives on cube rhombellane functionalized homeomorphs. *Studia Universitatis Babes-Bolyai Chemia* **2018**, *63*, 7–18. [CrossRef]
17. Diudea, M.V.; Medeleanu, M.; Khalaj, Z.; Ashrafi, A.R. Spongy Diamond. *Iran. J. Math. Chem.* **2019**, *10*, 1–9.
18. Trott, O.; Olson, A.J. AutoDockVina: Improving the speed and accuracy of docking with a new scoring function, efficient optimization and multithreading. *J. Comp. Chem.* **2010**, *31*, 455–461.
19. Bertrand, J.A.; Thieffine, S.; Vulpetti, A.; Cristiani, C.; Valsasina, B.; Knapp, S.; Kalisz, H.M.; Flocco, M. Structural characterization of the GSK-3beta active site using selective and non-selective ATP-mimetic inhibitors. *J. Mol. Biol.* **2003**, *333*, 393–407. [CrossRef]
20. Shoichet, B.K.; Kuntz, I.D.; Bodian, D.L. Molecular docking using shape descriptors. *J. Comput. Chem.* **2004**, *13*, 380–397. [CrossRef]
21. Dhananjayan, K.; Kalathil, K.; Sumathy, A.; Sivanandy, P. A computational study on binding affinity of bio-flavonoids on the crystal structure of 3-hydroxy-3-methyl-glutaryl-CoA reductase—An insilico molecular docking approach. *Der Pharma Chemica* **2014**, *6*, 378–387.
22. Abagyan, R.; Totrov, M. High-throughput docking for lead generation. *Curr. Opin. Chem. Biol.* **2001**, *5*, 375–382. [CrossRef]
23. Shen, M.; Zhou, S.; Li, Y.; Pan, P.; Zhang, L.; Hou, T. Discovery and optimization of triazine derivatives as ROCK1 inhibitors: Molecular docking, molecular dynamics simulations and free energy calculations. *Mol. Biosyst.* **2013**, *9*, 361–374. [CrossRef] [PubMed]

24. Shen, M.; Yu, H.; Li, Y.; Li, P.; Pan, P.; Zhou, S.; Zhang, L.; Li, S.; Lee, S.M.Y.; Hou, T. Discovery of Rho-kinase inhibitors by docking-based virtual screening. *Mol. Biosyst.* **2013**, *9*, 1511–1521. [CrossRef] [PubMed]
25. Jeffrey, G.A. *An Introduction to Hydrogen Bonding*; Oxford University Press: New York, NY, USA, 1997; Volume 12, p. 228.

Article

Docking of Polyethylenimines Derivatives on Cube Rhombellane Functionalized Homeomorphs

Beata Szefler * and Przemysław Czeleń

Department of Physical Chemistry, Faculty of Pharmacy, Collegium Medicum, Nicolaus Copernicus University, Kurpińskiego 5, 85-096 Bydgoszcz, Poland
* Correspondence: beatas@cm.umk.pl

Received: 22 July 2019; Accepted: 7 August 2019; Published: 14 August 2019

Abstract: Nowadays, in the world of science, an important goal is to create new nanostructures that may act as potential drug carriers. Among different, real or hypothetical, polymeric networks, rhombellanes are very promising and, therefore, attempts were made to deposit polyethylenimines as possible nano-drug complexes on the cube rhombellane homeomorphs surface. For the search of ligand–fullerene interactions, was used AutoDockVina software. As a reference structure, the fullerene C_{60} was used. After the docking procedure, the ligands–fullerenes interactions were tested. The important factor determining the mutual affinity of the tested ligands and nanocarriers is the symmetry of the analyzed nanostructures. Here, this feature has the influence on the distribution of such groups like donors and acceptors of hydrogen bonds on the surface of nanoparticles. We calculated the best binding affinities of ligands, values of binding constants and differences relative to C_{60} molecules. The best binding efficiency was found for linear ligands. It was also found that the shorter the molecule, the better the binding performance, the more the particle grows and the lower the yield. Small structures of ligands react easily with small structures of nanoparticles. The highest positive percentage deviations were obtained for ligand–fullerene complexes showing the highest binding energy values. Detailed analysis of structural properties after docking showed that the values of affinity of the studied indolizine ligands to the rhombellanes surface are correlated with the strength/length of hydrogen bonds formed between them.

Keywords: cube rhombellane homeomorph; PEI; polyethylenimines; nanostructure; molecular docking; affinity

---

1. Introduction

The development of multifunctional nanoparticles has a huge impact on the future of personalized medicine. Nanoparticles can be used for therapeutic purposes in anti-cancer therapy at the molecular level, which has been difficult so far. In earlier work [1–3], an enzyme GOx (3QVR) that fulfills the role of a biosensor, had been used for immobilization on the gel, while the polymer was polyethylenimine (PEI). By assembling polymeric nano-gels (for example PEI) and antibodies on nano-molecules [1–3], it was possible to recognize receptors of certain integrins on lung cancer tissues and to identify new cancer vessels.

In the present article, polyethylenimines (PEI) were studied. Molecular docking analysis of fifteen PEI derivatives acting as ligands on some cube rhombellane homeomorphs was carried out for the first time. Fourteen types of cube rhombellanes and three groups of polyethylenimines (PEIs), namely, branched (B-PEI), linear (L-PEI) and dendrimer (D-PEI) were used.

The choice of ligands and docked nanostructures was guided by our earlier studies [1–8].

PEIs (polyethylenimines) are polymeric molecules built of two aliphatic carbons and repeating units of amine groups. There are L-, B-, and DPEI (Figure 1). Linear PEI (LPEI) are built of secondary and primary amino groups (Figure 1, left); branched PEI (B-PEI) are built of all types amino groups such as primary, secondary and tertiary (Figure 1, middle), while dendrimers [9,10] are symmetric around the core (Figure 1, right). PEI despite the fact that are cytotoxic [11], have many applications, first of all, as transfection reagents [12].

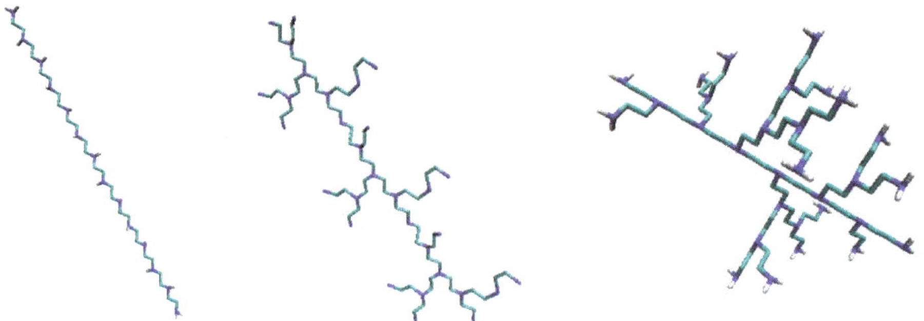

Figure 1. Examples of PEI molecules: linear L (**left**); branched B (**middle**) and dendrimer D (**right**) [13–15].

Calculations at the B3LYP/6-31G (d, p) level of theory [16–19] confirmed the hypothesis that rhombellanes are energetically feasible in the hope of a real synthesis [20–26].

Rhombellanes have certain specific traits which define these group of structures. At first, all strong rings are squares/rhombs. The second vertex classes consist of only non-connected vertices. Omega polynomial has a single term: $1X^{\wedge}|E|$ and they contain one K2.3 complete bipartite subgraph or the smallest rhombellane rbl.5. The end line graph of the parent graph has a Hamiltonian circuit.

To explore the internal molecular mobility that is important in the bioactivity study of these compounds we used the Molecular mechanics (MMFF94) [27] method. In this way we study the pharmaceutical important parameters of rhombellanes [27].

Rhombellans (Figure 2) seem to be structures suitable for medical chemistry, with a new class of structures which could be an used in personalized medicine as new carrier nanostructures.

Because rhombellane homeomorphs may be bound to a protein, an attempt was made to deposit PEI derivatives on rhombellanes, as possible nano-drug complexes. Detailed analysis of structural properties after docking showed many interesting features. Behavior of polyethylenimine (linear LPEI, branched BPEI and/or dendrimers DPEI) with respect to rhombellane homeomorphs, in terms of their (interacting) topology, geometry and energy, was studied. After the docking procedure, the best values of ligand–rhombellane affinity were found, which is an important result for homeomorphs.

The article is a collection of new data in the new field of rhombellanes (Figure 2).

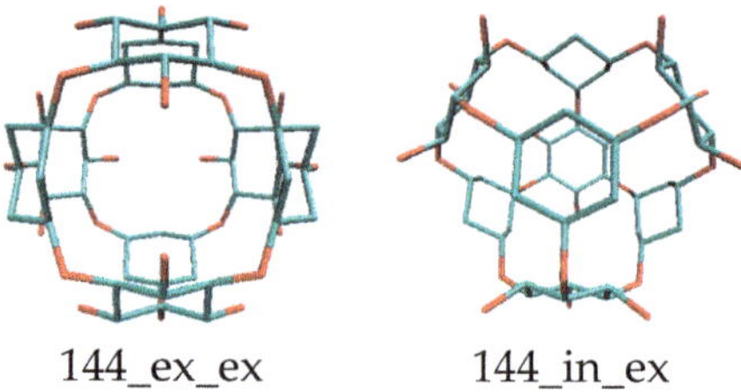

Figure 2. *Cont.*

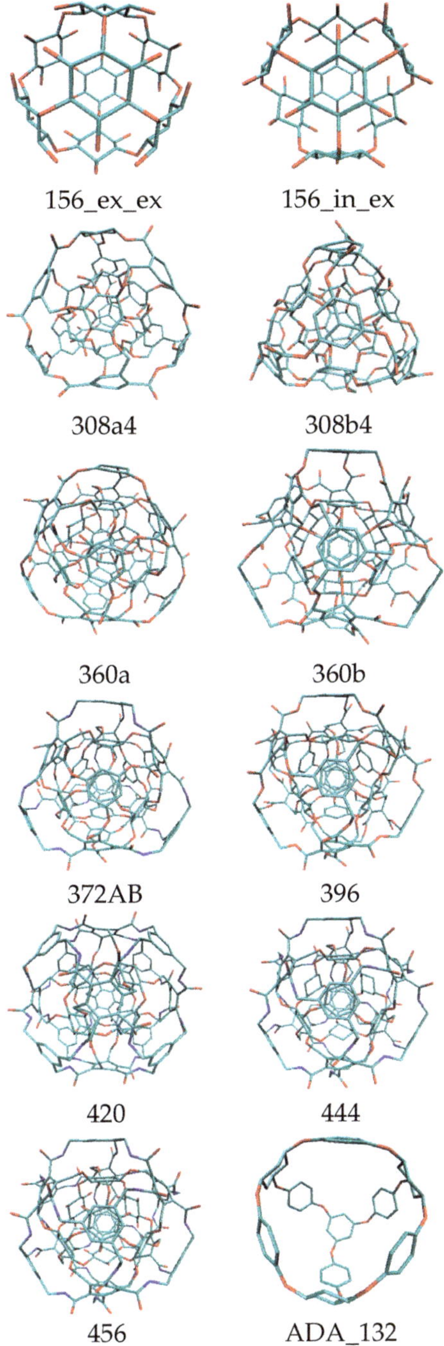

Figure 2. Graphic representation of the cube rhombellanes.

2. Methods

Docking Procedure

The ligand molecule was obtained from others study [1–8,13–15,28–31] Rhombellane homeomorphs, were received from Topo Cluj Group [32], the C60 structure was downloaded from Brookhaven Protein Database PDB [33], while C60 functionalized derivatives were obtained from PubChem database [34].

During the docking stage, there were structures of ligand and nano-systems containing only polar hydrogen atoms. In the case of all nanoparticles, the grid box dimensions were established equal to 26 × 26 × 26 Å, boxing coordinates amount to (0,0,0). All initial procedures related with preparation of ligand and nano-systems during the docking procedure were realized with the use of the AutoDock Tools package [35]. Using the AutoDockVina software in the docking procedure, after assigning hydrogen bonds, the molecules were loaded and stored as pdb-files [36]. The investigated ligands were loaded and their torsions along the rotatable bonds were assigned and next saved as "ligand.pdbqt". The grid menu after loading "pdbqt" was toggled [37]. For the search of ligand–rbl fullerene interactions, the map files were selected directly with setting up the grid points separately for each structure. The Lamarckian genetic algorithm completed the docking parameter files [38]. As a reference structure, the fullerene C60 was used, the most referred to structure in nanoscience.

All calculations during the docking stage were realized with the exhaustiveness parameter equal to 20, since such a value ensures an appropriate reproducibility of the results and a reasonable time of calculation. The structural analysis of considered systems and visualization of obtained complexes were realized with use of the VMD package [39]. The value of binding constant was calculated based on the formula:

$$K_{max} = \exp^{\left(\frac{-\Delta G_{max}}{RT}\right)}$$

where ΔG max represent maximal value of binding affinity obtained during docking stage, R represent value of gas constant and T temperature.

3. Results and Discussion

The results are presented in the following tables and figures. Rhombellane structures are given by their atom number.

In relation to the ligands of the D and L groups, the largest affinity values of the ligand–fullerenes were found for all ligands from B group (Figure 3, Table 1), for which the affinity values range from −2 to −7 kcal/mol. With the B_1320_PEI_C60N31 ligand the values are the lowest, thus showing the best affinity for all proposed fullerenes, with affinity values from −4 to −7 kcal/mol. All values of the interactions were compared with the values for C_{60} fullerene, which was, as always, used as the reference structure in nanostructures family. Therefore, in all cases of ligands from the B group, there are affinities with better and worse values of energy compared with affinity ligand–fulleren C_{60} (Figure 3, Table 1).

Table 1. The best binding affinity of ligands, BPEI, with the active site of Rbl-nano-structures (first column) during nine conformations.

	Gibbs Free Energy (kcal/mol)				
NANO-STRUCTURES	B10230_PEI_C22N12	B10230_PEI_C44N23	B10230_PEI_C20N11	B10230_PEI_C40N21	B10230_PEI_C60N31
144_ex_ex	−2.4	−2.9	−2.4	−2.8	−4.4
144_in_ex	−2.3	x	−2.2	−2.7	−4.3
156_ex_ex	−2.4	−2.8	−2.5	−2.6	−4.5
156_in_ex	−2.6	−2.9	−2.6	−2.9	−4.8
308a4	−3.9	−4.8	−3.9	−4.6	−6.4
308b4	−4.1	−5.0	−4.2	−4.6	−6.9
360a	−4.0	−4.5	−3.9	−4.5	−6.5
360b	−3.9	−4.5	−3.7	−4.2	−6.1

Table 1. *Cont.*

NANO-STRUCTURES	Gibbs Free Energy (kcal/mol)				
	B10230_PEI_C22N12	B10230_PEI_C44N23	B10230_PEI_C20N11	B10230_PEI_C40N21	B10230_PEI_C60N31
372AB	−4.3	−4.8	−4.0	−4.4	−6.2
396	−3.9	−4.7	−3.9	−4.2	−5.8
420	−4.2	−4.7	−3.7	−4.1	−5.4
444	−3.7	−4.3	−3.6	−4.0	−5.6
456	−3.9	−4.5	−3.5	−4.0	−5.3
ADA_132	−5.0	−4.4	−4.4	−4.2	x
C60	−3.8	−4	−3.8	−3.8	−4.8

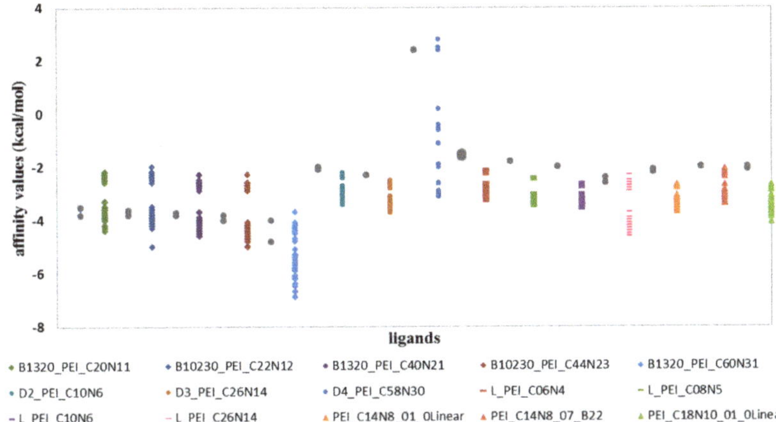

Figure 3. Energy (in kcal/mol) of the affinity ligand (PEI)-Rhombellanes complex (left and right), for B-, D-, and L-PEI; in scale to O kcal/mol.

When comparing the values of affinity of D ligand–fulleren and L ligand–fulleren complexes, it is clear that these complexes show the lowest values with all study fullerenes, compared with fulleren C$_{60}$ (Figure 3, Tables 2 and 3).

Table 2. The best binding affinity of ligands, BPEI, with the active site of Rbl-nano-structures (first column) during nine conformations.

NANO-STRUCTURES	Docked Energy (kcal/mol)		
	D2_PEI_C10N6	D3_PEI_C26N14	D4_PEI_C58N30
144_ex_ex	−2.4	−2.6	−2.6
144_in_e	−2.4	−2.8	−2.6
156_ex_ex	−3.1	−3.3	−3
156_in_ex	−3.1	−3.4	−3.1
308a4	−3.1	−3.6	−0.6
308b4	−3.3	−3.7	−2
360a	−3.1	−3.6	−2.9
360b	−2.9	−3.3	2.5
372AB	−3.2	−3.7	−1.1
396	−2.8	−3.6	4.9
420	−2.9	−3.5	5
444	−2.8	−3.3	10.4
456	−2.9	−3.3	8.2
ADA_132	−3.4	−2.7	−1.9
C$_{60}$	−2.1	−2.3	2.4

Table 3. The best binding energy of ligands, BPEI, with the active site of Rbl-nano-structures (first column) during nine conformations. In letters ABCD, etc., have been marked further nanostructures A-L_PEI_C06N4; B-L_PEI_C08N5; C-L_PEI_C10N6; D-L_PEI_C26N14; E-PEI_C14N8_01_0Linear; F-PEI_C14N8_07_B22; G-PEI_C18N10_01_0Linear.

NANO-STRUCTURES	Gibbs Free Energy (kcal/mol)						
	A	B	C	D	E	F	G
144_ex_ex	−2.3	−2.5	−2.8	−2.8	−2.8	−2.3	−2.9
144_in_ex	−2.2	−2.5	−2.8	−2.7	−2.8	−2.4	−2.9
156_ex_ex	−2.8	−3.2	−3.4	−2.6	−3.6	−2.9	−3.7
156_in_ex	−2.9	−3.3	−3.5	−2.9	−3.7	−3.2	−3.6
308a4	−3.1	−3.1	−3.4	−4.6	−3.4	−3.1	−3.8
308b4	−3.2	−3.5	−3.4	−4.6	−3.7	−3.4	−4.1
360a	−2.6	−3.1	−3.2	−4.5	−3.6	−3.2	−3.5
360b	−2.8	−3.2	−3.4	−4.2	−3.5	−3.2	−3.5
372AB	−2.7	−3.2	−3.3	−4.4	−3.5	−3.2	−3.8
396	−2.7	−3.2	−3.4	−4.2	−3.4	−3.1	−3.6
420	−2.6	−3.1	−3.4	−4.1	−3.5	−2.9	−3.5
444	−2.8	−3.3	−3.4	−4.0	−3.6	−3.1	−3.4
456	−2.8	−3.3	−3.5	−4	−3.6	−3	−3.6
ADA_132	−3.3	−3.5	−3.5	−4.2	−3.4	−3.4	−3.3
C_{60}	−1.6	−1.8	−2	−2.6	−2.2	−2	−2.1

For the other proposed ligands of type D, L and PEI_C14N8_01_Linear; PEI_C14N8_07_B22; PEI_C18N10_01_0, their affinities ranged from −2 to −4 kcal/mol (Tables 2 and 3). However, the ligand D4_PEI_C58N30 shows not only affinity in this range, i.e., (−2 to −4 kcal/mol), but also low affinities for fullerenes with high values of around 0 kcal/mol, and it does not even have the possibility of interacting with fullerenes as their affinity values are positive (Figure 3, Tables 2 and 3)

The diagram (Figure 4) shows two populations of affinity values of ligand B–fullerenes. The first with values ranging from −2 to −3 kcal/mol and in the case of second population from −3.3 to −5 kcal/mol for B1320_PEI_C20N11; B10230_PEI_C22N12; B1320_PEI_C40N21; B10230_PEI_C44N23.

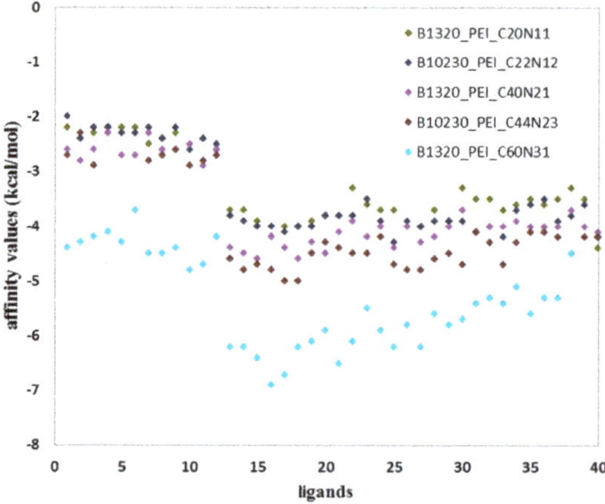

Figure 4. Example of diagram showing two populations of affinity values of ligand B–fullerenes.

Similarly, two populations are visible in the case of B1320_PEI_C60N31 with markedly reduced affinity values relative to the values of the affinity represented by the first four ligands of the B group, described above (Tables 3 and 4). The first population has interaction values ranging from −4 kcal/mol to −5 kcal/mol, the second shows much higher affinity values from −5 to −7 kcal/mol (Tables 3 and 4).

Table 4. The best binding affinity of ligands, BPEI, –columns A. Columns B represent k max values of binding constant estimated with use of binding free energy obtained for the best complex of ligand with nanostructure, while columns C represent k max differences relative C_{60} molecule, defining the difference in quality of ligand binding with considered nanostructure in comparison to reference system.

NANO-STRUCTURES	B10230_PEI C22N12			B10230_PEI C44N23			B10230_PEI C20N11			B10230_PEI C40N21			B10230_PEI C60N31		
	A	B	C	A	B	C	A	B	C	A	B	C	A	B	C
144_ex_ex	−2.4	57.4	−90.6	−2.9	133.6	−84.4	−2.4	57.4	−90.6	−2.8	112.8	−26.3	−4.4	1679.7	−8.3
144_in_ex	−2.3	48.5	−92.0		1.0	−99.9	−2.2	41.0	−93.3	−2.7	95.3	−28.9	−4.3	1418.9	−10.4
156_ex_ex	−2.4	57.4	−90.6	−2.8	112.8	−86.8	−2.5	68.0	−88.9	−2.6	80.5	−31.6	−4.5	1988.6	−6.3
156_in_ex	−2.6	80.5	−86.8	−2.9	133.6	−84.4	−2.6	80.5	−86.8	−2.9	133.6	−23.7	−4.8	3299.5	0.0
308a4	−4.0	855.1	40.2	−4.6	2354.2	175.3	−3.9	722.3	18.4	−4.6	2354.2	21.1	−6.4	49,120.4	33.3
308b4	−4.1	1012.4	65.9	**−5.0**	**4624.3**	**440.8**	**−4.2**	**1198.5**	**96.4**	−4.6	2354.2	21.1	**−6.9**	**114,226.6**	**43.8**
360a	−4.0	855.1	40.2	−4.5	1988.6	132.5	−3.9	722.3	18.4	**−4.5**	**1988.6**	**18.4**	**−6.5**	**58,151.8**	**35.4**
360b	−3.9	722.3	18.4	−4.5	1988.6	132.5	−3.7	515.4	−15.5	−4.2	1198.5	10.5	−6.1	29,604.6	27.1
372AB	**−4.3**	**1418.9**	**132.5**	**−4.8**	**3299.5**	**285.8**	−4.0	855.1	40.2	−4.4	1679.7	15.8	−6.2	35,047.8	14.6
396	−3.9	722.3	18.4	−4.7	2787.1	225.9	−3.9	722.3	18.4	−4.2	1198.5	10.5	−5.8	17,842.5	29.2
420	−4.2	1198.5	96.4	−4.7	2787.1	225.9	−3.7	515.4	−15.5	−4.1	1012.4	7.9	−5.4	9083.5	20.8
444	−3.7	515.4	−15.5	−4.3	1418.9	65.9	−3.6	435.3	−28.6	−4.0	855.1	2.6	−5.6	12,730.8	12.5
456	−3.9	722.3	18.4	−4.5	1988.6	132.5	−3.5	367.7	−39.7	−4.0	855.1	5.3	−5.3	7672.8	16.7
ADA_132	**−5.0**	**4624.3**	**657.9**	−4.4	1679.7	96.4	**−4.4**	**1679.7**	**175.3**	−4.2	1198.5	5.3		1.0	−6.3
C60	−3.8	610.2	0.0	−4.0	855.1	0.0	−3.8	610.2	0.0	−3.8	610.2	7.9	−4.8	3299.5	−100.0

Two populations are also visible for other ligands from the D and L groups forming interactions with tested fullerenes. Quantitatively, the largest number of ligand–fullerene interactions is expressed by affinity values in the range of −3 to −4 kcal/mol, and this is a representative population. The second population is expressed by a small representation of the number of affinities with values ranging from −2 to −3 kcal/mol.

The existence of these two populations is closely related to the interaction with two fullerene groups. Small structures of ligands easily react with small structures of nanoparticles and vice versa.

Unfortunately, the energy parameter itself is insufficient. By using energy per quantity of carbon atoms (kcal/mol) parameter it is clearly visible that the elongation of carbon chain does not affect the binding efficiency, but only increases affinity (Figure 5).

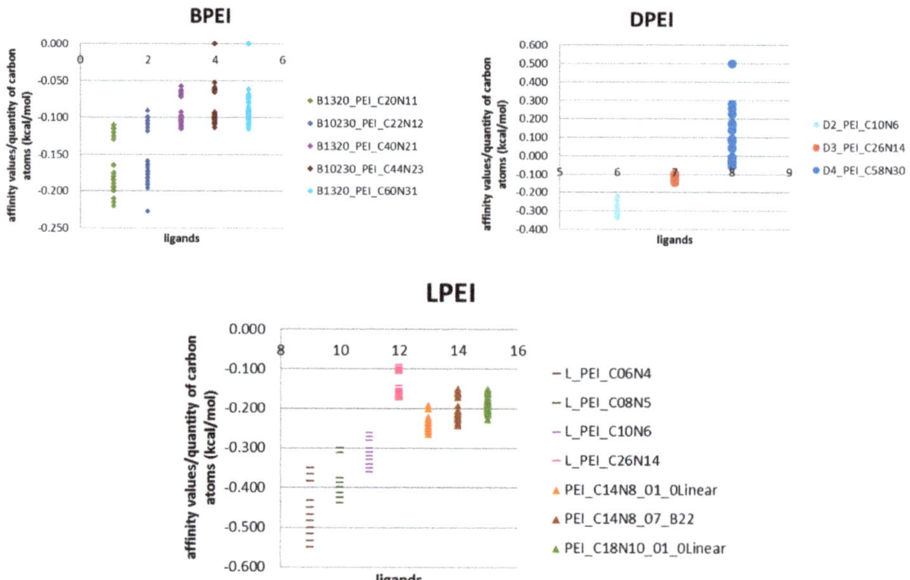

Figure 5. Affinity values per quantity of carbon atoms in kcal/mol for all ligands BPEI, DPEI and LPEI.

The best binding efficiency is shown by linear ligands L, with highest values of this parameter, compared with values of ligands from B and D groups (Figure 5). The shorter the molecule, the better the binding performance, the more the particle grows and the lower the yield. For linear LPEI structures, as the carbon chain length increases, the binding efficiency decreases and the saturation around the length of the chain with twenty carbon atoms is clearly visible (Figure 6).

Similar observations have been made for group B ligands. Twofold chain elongation results in a two-fold decrease in binding efficiency (Figure 5, Table 1). In the case of dendrimeric structures, as the complexity of the system increases, the value of binding efficiency decreases (Figure 5, Table 2). Among the fullerenes tested, the best effects were found for fullerenes ADA and 308a4/b4. Small ligands easily form complexes primarily with fullerene ADA, long-chain ligands interact with the 308a4/b4 nanostructure (Figure 5). Thus, for small ligands, the best binding efficiency is with small fullerenes, while large fullerenes require large ligands.

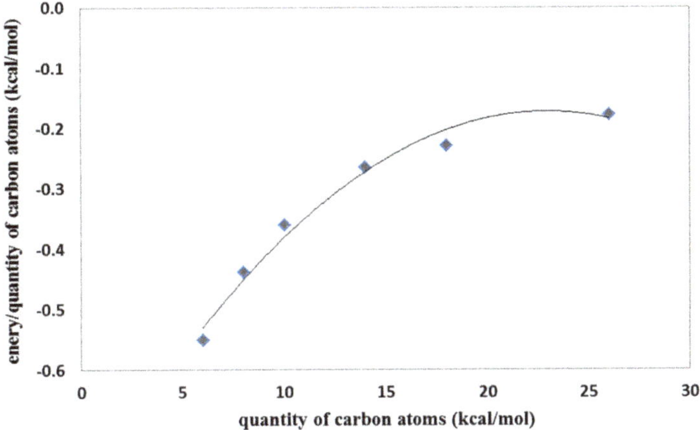

Figure 6. The affinity values per quantity of carbon atoms in function of quantity of carbon atoms for ligands PEI.

The best binding affinity of ligands BPEI (Table 4) DPEI (Table 5) and LPEI (Tables 6 and 7), k max values of binding constant estimated with use of binding free energy obtained for the best complex of ligand with nanostructure and k max differences relative C_{60} molecule, defining the difference in the quality of ligand binding with considered nanostructure in comparison to reference system, were estimated for the tested Rbl-structures in relation to the affinity value obtained for the fullerene C_{60}. The highest positive percentage deviations from the affinity of ligands to fullerene C_{60} were obtained for those Rbl-structures showing the highest binding values (Tables 4–7, in boldface). Two last columns show the equilibrium K value of the bonds.

Table 5. The best binding affinity of ligands, BPEI, –columns A. Columns B represent k max values of binding constant estimated with use of binding free energy obtained for the best complex of ligand with nanostructure, while columns C represent k max differences relative C_{60} molecule, defining the difference in quality of ligand binding with considered nanostructure in comparison to reference system.

NANO-STRUCTURES	D2_PEI_C10_N6			D3_PEI_C26_N14			D4_PEI_C58_N30		
	A	B	C	A	B	C	A	B	C
144_ex_ex	−2.4	57.44118	65.9216	−2.6	80.50545	65.9216	−2.6	80.50545	462,332.8
144_in_ex	−2.4	57.44118	65.9216	−2.8	112.8307	132.5439	−2.6	80.50545	462,332.8
156_ex_ex	−3.1	187.2105	440.7664	−3.3	262.3808	440.7664	−3	158.1354	908,248.5
156_in_ex	−3.1	187.2105	440.7664	−3.4	310.6226	540.1927	**−3.1**	**187.2105**	**1,075,259**
308a4	−3.1	187.2105	440.7664	−3.6	435.3464	797.2483	−0.6	2.752998	15,713.54
308b4	**−3.3**	**262.3808**	**657.8996**	**−3.7**	**515.39**	**962.2179**	−2	29.24283	167,874.3
360a	−3.1	187.2105	440.7664	**−3.7**	**515.39**	**962.2179**	**−2.9**	**133.5759**	**767,175.8**
360b	−2.9	133.5759	285.8405	−3.3	262.3808	440.7664	2.5	0.014705	−15.5307
372AB	−3.2	221.6313	540.1927	**−3.7**	**515.39**	**962.2179**	−1.1	6.401927	36,673.42
396	−2.8	112.8307	225.9168	−3.6	435.3464	797.2483	4.6	0.000425	−97.5601
420	−2.9	133.5759	285.8405	−3.5	367.7342	657.8996	5	0.000216	−98.7578
444	−2.8	112.8307	225.9168	−3.3	262.3808	440.7664	10.4	2.38E−08	−99.9999
456	−2.9	133.5759	285.8405	−3.3	262.3808	440.7664	8.2	9.76E−07	−99.9944
ADA_132	**−3.4**	**310.6226**	**797.2483**	−2.7	95.30732	96.42822	−1.9	24.70122	141,786.7
C60_2	−2.1	34.61947	0	−2.3	48.52017	0	2.4	0.017409	0

Table 6. The best binding affinity of ligands, BPEI, –columns A. Columns B represent k max values of binding constant estimated with use of binding free energy obtained for the best complex of ligand with nanostructure, while columns C represent k max differences relative C_{60} molecule, defining the difference in quality of ligand binding with considered nanostructure in comparison to reference system.

NANO-STRUCTURES	L_PEI_C06N4			L_PEI_C08N5			L_PEI_C10N6			L_PEI_C26N14		
	A	B	C	A	B	C	A	B	C	A	B	C
144_ex_ex	−2.3	48.5	225.9	−2.5	68.0	225.9	−2.8	112.8	285.	−2.8	112.8	40.2
144_in_ex	−2.2	41.0	175.3	−2.5	68.0	225.9	−2.8	112.8	285.8	−2.7	95.3	18.4
156_ex_ex	−2.8	112.8	657.9	−3.2	221.6	962.2	−3.4	310.6	962.2	−2.6	80.5	0.0
156_in_ex	−2.9	133.6	797.2	−3.3	262.4	1157.5	**−3.5**	**367.7**	1157.5	−2.9	133.6	65.9
308a4	−3.0	158.1	962.2	−3.3	262.4	1157.5	−3.4	310.6	962.2	**−4.6**	**2354.2**	**2824.3**
308b4	**−3.2**	**221.6**	**1388.7**	**−3.5**	**367.7**	**1662.4**	−3.4	310.6	962.2	**−4.6**	**2354.2**	**2824.3**
360a	−2.7	95.3	540.2	−3.1	187.2	797.2	−3.2	221.6	657.9	−4.5	1988.6	2370.1
360b	−2.8	112.8	657.9	−3.2	221.6	962.2	−3.4	310.6	962.2	−4.2	1198.5	1388.7
372AB	−2.7	95.3	540.2	−3.2	221.6	962.2	−3.4	310.6	962.2	−4.4	1679.7	1986.5
396	−2.7	95.3	540.2	−3.2	221.6	962.2	−3.4	310.6	962.2	−4.2	1198.5	1388.7
420	−2.7	95.3	540.2	−3.1	187.2	797.2	**−3.6**	**435.3**	**1388.7**	−4.1	1012.4	1157.5
444	−2.8	112.8	657.9	−3.3	262.4	1157.5	−3.4	310.6	962.2	−4.0	855.1	962.2
456	−2.8	112.8	657.9	−3.3	262.4	1157.5	**−3.5**	**367.7**	**1157.5**	−4.0	855.1	962.2
ADA_132	**−3.3**	**262.4**	**1662.4**	**−3.5**	**367.7**	**1662.4**	**−3.5**	**367.7**	1157.5	−4.2	1198.5	1388.7
C60	−1.6	14.9	0.0	−1.8	20.9	0.0	−2.0	29.2	0.0	−2.6	80.5	0.0

Table 7. The best binding affinity of ligands, BPEI, –columns A. Columns B represent k max values of binding constant estimated with use of binding free energy obtained for the best complex of ligand with nanostructure, while columns C represent k max differences relative C_{60} molecule, defining the difference in quality of ligand binding with considered nanostructure in comparison to reference system.

NANO-STRUCTURES	PEI_C14N8_01_0Linear			PEI_C14N8_07_B22			PEI_C18N10_01_0Linear		
	A	B	C	A	B	C	A	B	C
144_ex_ex	−2.8	112.8307	40.15285	−2.3	48.52017	175.2998	−2.9	133.5759	37.5
144_in_ex	−2.8	112.8307	65.9216	−2.4	57.44118	225.9168	−2.9	133.5759	37.5
156_ex_ex	−3.6	435.3464	225.9168	−2.9	133.5759	657.8996	−3.7	515.39	62.5
156_in_ex	**−3.7**	**515.39**	**657.8996**	−3.2	221.6313	1157.519	−3.7	515.39	75
308a4	−3.2	221.6313	440.7664	−3.1	187.2105	1388.729	−3.7	515.39	87.5
308b4	**−3.7**	**515.39**	**540.1927**	**−3.4**	**310.6226**	**2824.283**	**−4.1**	**1012.371**	**93.75**
360a	−3.6	435.3464	440.7664	−3.2	221.6313	962.2179	−3.5	367.7342	62.5
360b	−3.5	367.7342	356.7818	−3.2	221.6313	797.2483	−3.5	367.7342	68.75
372AB	−3.5	367.7342	657.8996	−3.2	221.6313	1662.449	**−3.8**	**610.1504**	68.75
396	−3.4	310.6226	440.7664	−3.1	187.2105	1157.519	−3.6	435.3464	68.75
420	−3.5	367.7342	440.7664	−3	158.1354	962.2179	−3.5	367.7342	68.75
444	−3.6	435.3464	356.7818	−3.1	187.2105	797.2483	−3.4	310.6226	75
456	−3.6	435.3464	440.7664	−3	158.1354	1157.519	−3.6	435.3464	68.75
ADA_132	−3.4	310.6226	797.2483	**−3.4**	**310.6226**	440.7664	−3.3	262.3808	100
C60	−2.2	40.98466	0	−2	29.24283	0	−2.1	34.61947	0

The higher the K value, the more the reaction proceeds towards the formation of the complex.

Detailed analysis of structural properties after docking showed that the affinities of the ligands to the rhombellanes surface are correlated with the quality of hydrogen bonds formed between them. The distance between acceptor and hydrogen atoms is the criterion for classification of the strength of hydrogen bonds: weak interactions are characterized by distances <3 Å, strong interactions by a distance <1.6 Å and medium strength by values in the range from 1.6 Å to 2.0 Å.

Ligand B1320_PEI_C20N11 and nanostructure ADA_132 form two hydrogen bonds with medium strength between amino groups of ligand and oxygen atoms of fullerenes with binding lengths 2.79Å and 2.95Å (Figure 7). Also, in the case of ligand B1320_PEI_C60N31–fullerene 308b4 four medium hydrogen bonds were created with bond lengths 2.69 Å, 2.87 Å and 2.93 Å (Figure 7).

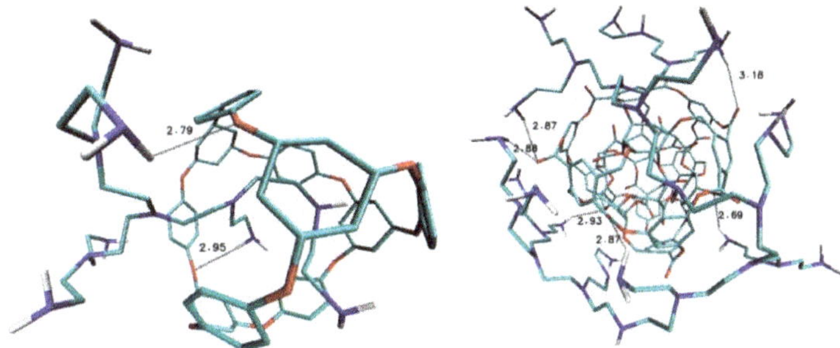

Figure 7. Interactions found in the complexes of fullerene ADA_132 and ligands B1320_PEI_C20N11 (**left**) and fullerene 308b4 with ligand B1320_PEI_C60N31 (**right**) after the docking procedure.

After the docking of ligands from D group, different interactions could be found, namely in the case of fullerene ADA_132 and ligands D2_PEI_C10N6 there is only one week hydrogen bond, while for fullerene 308b4 with D3_PEI_C26N14 ligand there are several strong and medium hydrogen bonds, first of all between amino groups of ligand and oxygen atom of nanostructure with bond lengths 1.9 Å, 2.47 Å., 2.52 Å and 2.94 Å. In the case of fullerene 156_in_ex with ligand D4_PEI_C58N30, there are three hydrogen bonds of medium strength; with bond lengths 2.50 Å, 2.80 Å and 2.83 Å (Figure 8).

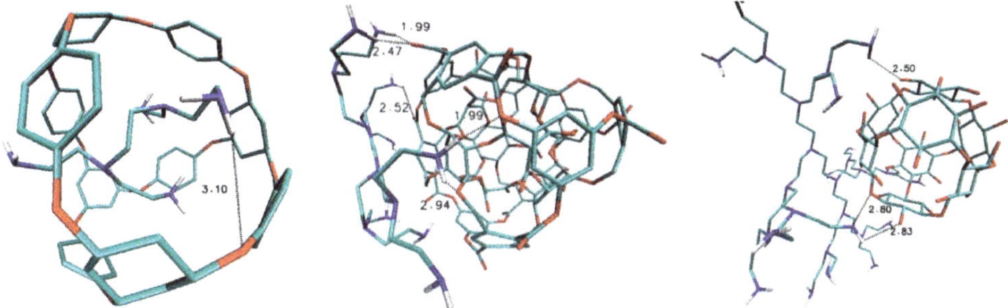

Figure 8. Interactions found in the complexes of fullerene ADA_132 and ligands D2_PEI_C10N6 (**left**) and fullerene 308b4 with D3_PEI_C26N14 ligand (**middle**) and 156_in_ex with ligand D4_PEI_C58N30 (**right**) after the docking procedure.

After the docking of ligands from L group, different interactions were also found, namely in the case of fullerene ADA_132 and ligands L_PEI_C10N6 there is only one-week hydrogen bond, the same as in the case D2_PEI_C10N6-ADA_132 (Figure 9). Again, as in DPEI-308b4 case, there are many interactions between hydrogen atoms of nitrogen groups of ligand and oxygen atoms of nanostructure with values 1.99 Å, 2.47 Å, 2.52 Å, 2.94 Å and 3.05 Å (Figure 9).

In the case of fullerene 308b4 and ligand PEI_C18N10_01_0Linear, there are several strong and medium hydrogen bonds, while in the case of fullerene C_{60} as references structure with B1320_PEI_C60N31 ligand there are no important interactions (Figure 10).

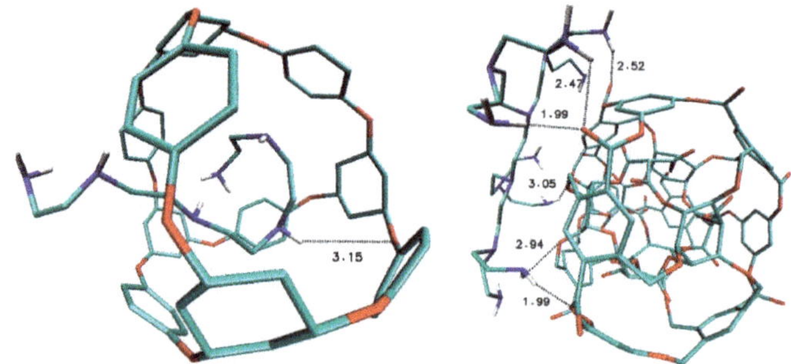

Figure 9. Interactions found in the complexes of fullerene ADA_132 and ligands L_PEI_C10N6 (**left**) and fullerene 308b4 with ligand L_PEI_C26N14 (**right**) after the docking procedure.

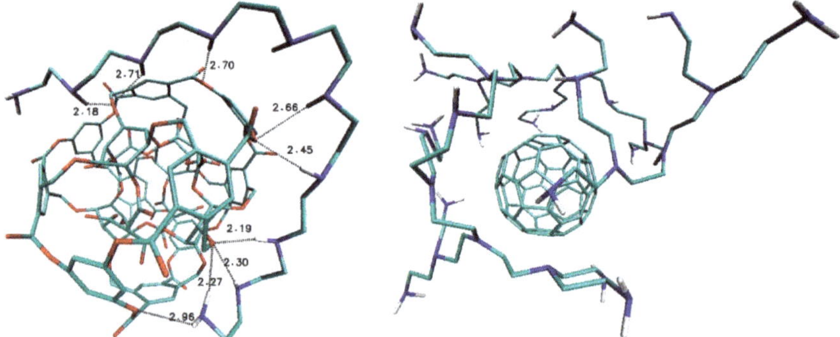

Figure 10. Interactions found in the complexes of fullerene 308b4 and ligands PEI_C18N10_01_0Linear (**left**) and fullerene c60 as references structure with ligand B1320_PEI_C60N31 (**right**) after the docking procedure.

4. Conclusions

As a proposal for a new nanodrug, an attempt was made to implement PEI ligands on the cube rhombellane homeomorphic surface. Fourteen types of cube rhombellanes were used together with three groups of polyethylenimines (PEIs), namely, branched (B-PEI), linear (L-PEI) and dendrimer (D-PEI). Ligand-fullerenes interactions were described in terms of quality and quantity. Specifically, there were calculated the affinity values and affinity per quantity of carbon atoms after the docking procedure for ligand nanostructure. The best binding efficiency was shown by linear ligands L, with highest values of this parameter, compared with values of ligand from B and D groups. The shorter the molecule, the better the binding performance, the more the particle grows and the lower the yield. For linear structures LPEI, as the carbon chain length increases, the binding efficiency decreases and the saturation around the length of the chain with twenty carbon atoms is clearly visible. Similar observations have been made for group B ligands. Twofold chain elongation results in a two-fold decrease in binding efficiency. In the case of dendrimeric structures, as the complexity of the system increases, the value of binding efficiency decreases. Two populations of affinity values have been observed, which is closely related to the interaction with two fullerene groups. Small structures of ligands easily react with small structures of nanoparticles and vice versa. The best binding affinity of ligands and k max values of binding constant were estimated with the use of binding free energy obtained for the best complex of ligand with nanostructure. Also, k max differences relative C_{60} molecule, defining the difference in quality of ligand binding with considered nanostructure in comparison to reference system, were calculated. The highest positive percentage deviations were obtained for ligand–fullerene complexes showing the highest binding energy values. Detailed analysis of structural properties after docking showed that the values of affinity of the studied indolizine ligands to the Rhombellanes surface are correlated with the strength/length of hydrogen bonds formed between them.

Author Contributions: Conceptualization, B.S.; Methodology, B.S. and P.C.; Validation, B.S. and P.C.; Formal Analysis, B.S.; Investigation, B.S.; Resources, B.S.; Data Curation, B.S.; Writing—Original Draft Preparation, B.S.; Writing—Review & Editing, B.S. and P.C.; Visualization, B.S. and P.C.; Supervision, B.S.; Project Administration, B.S.; Funding Acquisition, B.S.

Funding: This research received no external funding.

Acknowledgments: Thanks for fruitful cooperation for many years to MV Diudea; this article was supported by PL-Grid Infrastructure (http://www.plgrid.pl/en).

Conflicts of Interest: The authors declare no conflict of interest.

References and Note

1. Lungu, C.L.; Diudea, M.V.; Putz, M.V.; Grudzinski, I.P. Linear and Branched PEIs (Polyethylenimines) and Their Property Space. *Int. J. Mol. Sci.* **2016**, *17*, 555. [CrossRef]
2. Szefler, B.; Diudea, M.V.; Grudziński, I.P. Nature of Polyethyleneimine-Glucose Oxidase interactions. *Stud. Univ. Babes-Bolyai Chem.* **2016**, *61*, 249–260.
3. Szefler, B.; Diudea, M.V.; Putz, M.V.; Grudziński, I.P. Molecular Dynamic Studies of the Complex Polyethylenimine and Glucose Oxidase. *Int. J. Mol. Sci.* **2016**, *17*, 1796. [CrossRef]
4. Szefler, B.; Czeleń, P.; Diudea, M.V. Docking of Indolizine derivatives on cube Rhombellane functionalized Homeomorphs. *Stud. Univ. Babes-Bolyai Chem.* **2018**, *63*, 7–18. [CrossRef]
5. Czeleń, P. Investigation of the Inhibition Potential of New Oxindole Derivatives and Assessment of Their Usefulness for Targeted Therapy. *Symmetry* **2019**, *11*, 974. [CrossRef]
6. Czeleń, P.; Szefler, B. The Immobilization of ChEMBL474807 Molecules Using Different Classes of Nanostructures. *Symmetry* **2019**, *11*, 980. [CrossRef]
7. Czeleń, P.; Szefler, B. The Immobilization of Oxindole Derivatives with Use of Cube Rhombellane Homeomorphs. *Symmetry* **2019**, *11*, 900. [CrossRef]
8. Szefler, B.; Czeleń, P. Docking of Cisplatin on Fullerene Derivatives and Some Cube Rhombellane Functionalized Homeomorphs. *Symmetry* **2019**, *11*, 874. [CrossRef]
9. Astruc, D.; Boisselier, E.; Ornelas, C. Dendrimers Designed for Functions: From Physical, Photophysical, and Supramolecular Properties to Applications in Sensing, Catalysis, Molecular Electronics, and Nanomedicine. *Chem. Rev.* **2010**, *110*, 1857–1959. [CrossRef]
10. Vögtle, F.; Richardt, G.; Werner, N. *Dendrimer Chemistry Concepts, Syntheses, Properties, Applications*; John Wiley & Sons: Hoboken, NJ, USA, 2009; ISBN 3-527-32066-0.
11. Akinc, A.; Thomas, M.; Klibanov, A.M.; Langer, R. Exploring polyethylenimine-mediated DNA transfection and the proton sponge hypothesis. *J. Gene Med.* **2004**, *7*, 657–666. [CrossRef]
12. Rudolph, C.; Lausier, J.; Naundorf, S.; Müller, R.H.; Rosenecker, J. In vivo gene delivery to the lung usingpolyethylenimine and fractured polyamidoamine dendrimers. *J. Gene Med.* **2000**, *2*, 269–278. [CrossRef]
13. Poplawska, M.; Bystrzejewski, M.; Grudzinski, I.P.; Cywinska, M.A.; Ostapko, J.; Cieszanowski, A. Immobilization of gamma globulins and polyclonal antibodies of class IgG onto carbon-encapsulated iron nanoparticles functionalized with various surface linkers. *CARBON* **2014**, *74*, 180–194. [CrossRef]
14. Karachevtsev, V.A.; Glamazda, A.Y.U.; Zarudnev, E.S.; Karachevtsev, M.V.; Leontiev, V.S.; Linnik, A.S.; Lytvyn, O.S.; Plokhotnichenko, A.M.; Stepanian, S.G. Glocose oxidase immobilization onto carbonnanotube networking. *Ukr. J. Phys.* **2012**, *57*, 700.
15. Kasprzak, A.; Popławska, M.; Bystrzejewski, M.; Łabędź, O.; Grudziński, I.P. Conjugation of polyethylenimine and its derivatives to carbon-encapsulated iron nanoparticles. *R. Soc. Chem. Adv.* **2015**, *5*, 85556. [CrossRef]
16. Gomes, J.A.N.F.; Mallion, R.B. Aromaticity and ring currents. *Chem. Rev.* **2001**, *101*, 1349–1383. [CrossRef]
17. Cyrański, M.K.; Krygowski, T.M.; Katritzky, A.R.; Schleyer, P.V.R. To what extent can aromaticity be defined uniquely? *J. Org. Chem.* **2002**, *67*, 1333–1338.

18. Chen, Z.; Wannere, C.S.; Crominboeuf, C.; Puchta, R.; Schleyer, R.V.P. Nucleus-independent chemical shifts (NICS) as an aromaticity criterion. *Chem. Rev.* **2005**, *105*, 3842–3888. [CrossRef]

19. Diudea, M.V.; Lungu, C.N.; Nagy, C.L. Cube-rhombellane related structures: A drug perspective. *Molecules* **2018**, *23*, 2533. [CrossRef]

20. Pauling, L.; Wheland, G.W. The nature of the chemical bond. V. The quantum mechanical calculation of the resonance energy of benzene and naphthalene and the hydrocarbon free radicals. *J. Chem. Phys.* **1933**, *1*, 362–374. [CrossRef]

21. Daudel, R.; Lefebre, R.; Moser, C. *Quantum Chemistry*; Interscienc: New York, NY, USA, 1959.

22. Diudea, M.V.; Cataldo, F. Functionalized rhombellanes in Drug Design. *Curr. Comput.-Aided Drug Des.* **2018**. accepted.

23. Pop, R.; Medeleanu, M.; Diudea, M.V.; Szefler, B.; Cioslowski, J. Fullerenes patched by flowers. *Cent. Eur. J. Chem.* **2013**, *11*, 527–534. [CrossRef]

24. Frisch, M.J.; Trucks, G.W.; Schlegel, H.B.; Scuseria, G.E.; Robb, M.A.; Cheeseman, J.R.; Scalmani, G.; Barone, V.; Mennucci, B.; Petersson, G.A.; et al. *Gaussian 09, Revision, A.1*; Gaussian, Inc.: Wallingford, CT, USA, 2009.

25. Randić, M. Aromaticity of polycyclic conjugated hydrocarbons. *Chem. Rev.* **2003**, *103*, 3449–3605. [CrossRef] [PubMed]

26. Diudea, M.V.; Nagy, C.L. *Periodic Nanostructures*; Springer: Dordrecht, The Netherlands, 2007.

27. Pauling, L. *The Nature of the Chemical Bond University*; Cornell University Press: Ithaca, NY, USA, 1960; Volume 260.

28. Kasprzak, A.; Grudziński, I.P.; Bamburowicz-Klimkowska, M.; Parzonko, A.; Gawlak, M.; Poplawska, M. New insight into Synthesis and Biological Activity of the Polymeric Materials Consisting of Folic Acid and β-cyclodextrin. *Macromol. Biosci.* **2018**, *18*, 1700289. [CrossRef] [PubMed]

29. Kasprzak, A.; Poplawska, M.; Bystrzejewski, M.; Grudzinski, I.P. Sulfhydrylated graphene-encapsulated iron nanoparticles directly aminated with polyethylenimine: A novel magnetic nanoplatform for bioconjugation of gamma globulins and polyclonal antibodies. *J. Mater. Chem. B* **2016**, *4*, 5593. [CrossRef]

30. Bamburowicz-Klimkowska, M.; Poplawska, M.; Grudzinski, I.P. Nanocomposites as biomolecules delivery agents in nano-medicine. *J. Nanobiotechnol.* **2019**, *17*, 48. [CrossRef]

31. Kowalczyk, A.; Sęk, J.P.; Kasprzak, A.; Popławska, M.; Grudzinski, I.P.; Nowicka, A.M. Occlusion phenomenon of redox probe by protein asa way of voltammetric detection of non-electroactive C-reactive protein. *Biosens. Bioelectron.* **2018**, *117*, 232–239. [CrossRef] [PubMed]

32. Topo-Cluj—International scientific group created by prof. M.V. Diudea, Babes-Bolyai University, Faculty of Chemistry and Chemical Engineering, Cluj-Napoca, Romania

33. Kim, K.-H.; Ko, D.K.; Kim, Y.-T.; Kim, N.H.; Paul, J.; Zhang, S.-Q.; Murray, C.B.; Acharya, R.; Kim, Y.H.; DeGrado, W.F.; et al. Protein-directed self-assembly of a fullerene crystal. *Nat. Commun.* **2016**, *7*, 11429. [CrossRef] [PubMed]

34. PubChem. Available online: https://pubchem.ncbi.nlm.nih.gov/May (accessed on 12 May 2019).

35. Shoichet, B.K.; Kuntz, I.D.; Bodian, D.L. Molecular docking using shape descriptors. *J. Comput. Chem.* **2004**, *13*, 380–397. [CrossRef]

36. Trott, O.; Olson, A.J. AutoDock Vina: Improving the speed and accuracy of docking with a new scoring function, efficient optimization, and multithreading. *J. Comput. Chem.* **2010**, *31*, 455–461. [CrossRef]

37. Dhananjayan, K.; Kalathil, K.; Sumathy, A.; Sivanandy, P. A computational study on binding affinity of bio-flavonoids on the crystal structure of 3-hydroxy-3-methyl-glutaryl-CoA reductase—An insilico molecular docking approach. *Der Pharma Chem.* **2014**, *6*, 378–387.

38. Abagyan, R.; Totrov, M. High-throughput docking for lead generation. *Curr. Opin. Chem. Biol.* **2001**, *5*, 375. [CrossRef]

39. Humphrey, W.; Dalke, A.; Schulten, K. VMD: Visual molecular dynamics. *J. Mol. Graph.* **1996**, *14*, 33–38. [CrossRef]

Article

Computational Exploration of Functionalized Rhombellanes: Building Blocks and Double-Shell Structures

Katalin Nagy [1], Beata Szefler [2] and Csaba L. Nagy [1],*

[1] Department of Chemistry and Chemical Engineering, Faculty of Chemistry and Chemical Engineering, Babes-Bolyai University, Arany J. street 11, RO-400028 Cluj-Napoca, Romania; knagy@chem.ubbcluj.ro
[2] Department of Physical Chemistry, Faculty of Pharmacy, Collegium Medicum, Nicolaus Copernicus University, Kurpińskiego 5, 85-096 Bydgoszcz, Poland; beatas@cm.umk.pl
* Correspondence: nc35@chem.ubbcluj.ro

Received: 30 January 2020; Accepted: 18 February 2020; Published: 1 March 2020

Abstract: Double-shell covalent assemblies with the framework of the cube–rhombellane were recently proposed as potential drug delivery systems. Their potential to encapsulate guest molecules combined with appropriate surface modifications show great promise to meet the prerequisites of a drug carrier. This work reports the molecular design of such clusters with high molecular symmetry, as well as the evaluation of the geometric and electronic properties using density functional theory. The computational studies of the double-shell assemblies and their corresponding building blocks were conducted using the B3LYP/6-31G(d,p) method as implemented in Gaussian 09. The results show that the assembly of the building blocks is energetically favorable, leading to clusters with higher stability than the corresponding shell fragments, with large HOMO–LUMO gap values. In case of aromatic systems, interlayer stacking interactions between benzene rings contribute to the molecular geometry and stability. During geometry optimization the clusters preserve the high molecular symmetry of the building blocks.

Keywords: double-shell structures; DFT; cube–rhombellanes; covalent assembly; van der Waals interaction; drug delivery

---

1. Introduction

Over the past years great effort was dedicated to the synthesis of molecular assemblies which provide cavities that enable the encapsulation of guest molecules in a wide range of size scales. Cryptophanes [1] and carcerands [2] were the first cage-like molecular complexes based on covalent bonds. Subsequently, host molecules with space-restricted properties were assembled by metal–ligand interactions [3] or weak interactions like hydrogen bonds [4,5].

A particularly fascinating class of double- and multi-shell molecules is the spherical carbon nanostructures, known as onion or nested fullerenes. They consist of several concentric graphitic layers where a giant fullerene encapsulates progressively smaller cages. In such assemblies, weak nonbonding inter-shell interactions exist; therefore, the distance between adjacent shells plays a major role in the stability [6,7].

The hypothetical double-shell hydrocarbon called hyper-cubane [8] and its functionalized derivatives [9] were proposed and computationally investigated.

The recently proposed [10,11] cube–rhombellane **1a** is a double-shell structure, designed by graph–theoretical transformation called "rhombellation" of the cube [10], which serves as a framework for chemical structures with both high complexity and symmetry, hereafter called rhombellanes. The core of the framework is the cube, shown as green spheres in Figure 1. The surface layer of **1a** is composed of six hexavalent and eight trivalent vertices, shown as red and yellow spheres, respectively. The molecular realization of **1a** requires the appropriate structural fragments that can establish up to six connections with neighboring atoms. According to the purpose, six-fold rings, i.e., cyclohexane or benzene, were found to be suitable.

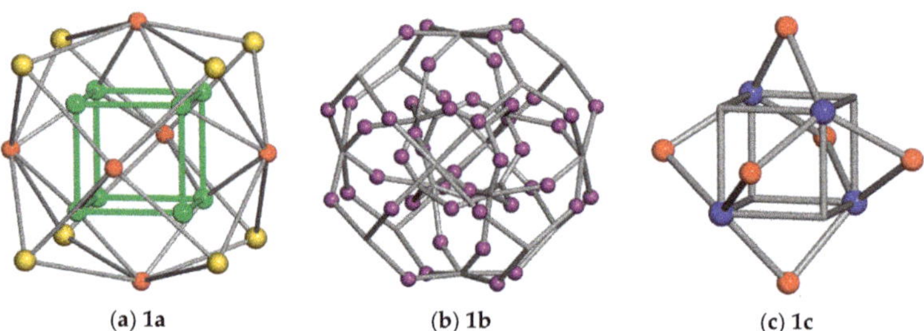

(a) 1a   (b) 1b   (c) 1c

Figure 1. The dual layer cube–rhombellane (**a**) and derived frameworks (**b**,**c**).

Structure **1b** represents the homeomorph of **1a**, where the violet spheres (Figure 1) represent molecular linkers for the covalent binding of the two layers. In structure **1c** the relative position of the four hexavalent vertices (blue spheres) in the inner shell of **1a** is highlighted.

Our previous computational results on rhombellanes [12] have shown that rhombellanes are potential alternatives as drug carriers, indicated by their ADME (absorption, distribution, metabolism, and excretion) properties. To serve as a drug delivery system the molecule should possess several prerequisites, including sufficiently strong adsorptive effects towards bioactive molecules, to ensure the delivery to the target site. Therefore, further studies were performed on rhombellanes to evaluate the immobilization potential of different organic compounds. Several rhombellanes showed satisfying binding affinities towards different ligand molecules, including indirubin derivatives (ChEMBL474807 molecule) [13], oxindole derivatives [14], and cisplatin [15], investigated by molecular docking methods. The results confirmed that the distribution of the hydrogen bond donors and acceptors on the surface of rhombellanes, as well as stacking interactions between aromatic systems of both molecules, significantly contribute to the binding capacity of such systems.

These findings further motivated us to perform a systematic study to find which building blocks are suitable for the assembly of rhombellanes, from both a geometric and stability point of view. This paper presents the computational investigation using density functional theory (DFT) of both the inner (core) and surface layers and the corresponding double-shell assemblies of some rhombellanes with high molecular symmetry. Although it is not explored in this work, another important feature of this class of compounds is their hollow inside, which enables them to encapsulate metal atoms or smaller guest molecules.

2. Methods

Ground state geometries and electronic properties of the discussed structures were obtained using density functional theory. Initial geometries were fully optimized using the hybrid density functional B3LYP and 6-31G(d,p) basis set, and the Cartesian coordinates of all molecules are provided in the Supplemental Materials. To ensure that optimized structures correspond to a stationary point, vibrational frequencies were computed at the same level of theory. All calculations were performed using the Gaussian 09 computational chemistry software package [16].

The initial geometries of the double-shell clusters converged during optimization only if the input geometry was built from the already geometry optimized layers, and most importantly the linkers between the layers were positioned to maintain the molecular symmetry. For the geometry optimization, tight convergence criteria and symmetry constraints were applied. However, even without symmetry constraints, the structures maintained their starting point group symmetry during geometry optimization.

3. Results and Discussion

3.1. Structural Models

In the present study, only core building blocks with C–O–C linkage between carbon rings (both benzene and cyclohexane) were selected. For the surface layer we limited the selection to molecules where the aromatic rings are connected by ester or amide chemical bonds. Although several double-shell clusters were designed from such fragments, in this work only the energetically feasible rhombellanes are presented. Building of a cluster with the framework of rhombellane requires that the surface layer can easily accommodate the core fragment, and also, they must be properly oriented such that linkers can covalently bind the two layers with the least geometric strain. Both fragments have high molecular symmetry; the inner layer has octahedral (O_h) or tetrahedral (T_d), whereas due to the amide and ester bonds the surface layer has tetrahedral (T) point group symmetry.

3.1.1. Core Building Blocks

The molecules **2a–2f** displayed in Figure 2 correspond to the core building blocks of the double-shell structures. Each structure included eight six-fold rings arranged at the vertices of a cube, which were linked by oxygen atoms, shown as red spheres in Figure 2. Structures **2a–2d** were built only from cyclohexane rings, where the linking oxygen atom was connected to either equatorial or axial positions. Structure **2e** consisted of eight benzene rings in an octahedral arrangement, whereas **2f** contained both aromatic and cycloaliphatic rings, which were aligned at the vertices of a tetrahedron. Notice that each cyclohexane ring adopted a chair conformation. With one exception, structure **2b**, the oxygen atoms pointed outwards from the carbon cluster.

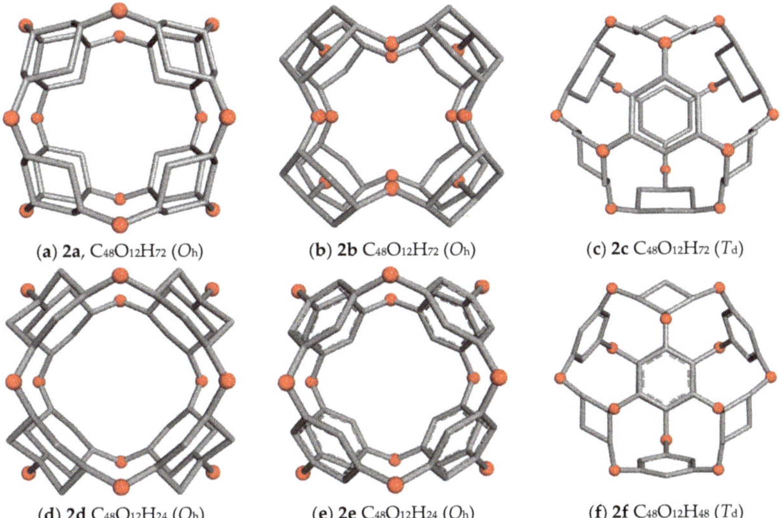

(**a**) **2a**, $C_{48}O_{12}H_{72}$ (O_h)   (**b**) **2b** $C_{48}O_{12}H_{72}$ (O_h)   (**c**) **2c** $C_{48}O_{12}H_{72}$ (T_d)

(**d**) **2d** $C_{48}O_{12}H_{24}$ (O_h)   (**e**) **2e** $C_{48}O_{12}H_{24}$ (O_h)   (**f**) **2f** $C_{48}O_{12}H_{48}$ (T_d)

Figure 2. Optimized geometries of the core fragments considered for the assembly of double-shell structures. To highlight the position of the aromatic and cyclohexane rings, the figure displays the perspective views of the molecules. Carbon rings are linked via an ether bond, and the red spheres correspond to the oxygen atoms. Delocalized bonds reveal the location of aromatic rings in **2e** and **2f**.

The aliphatic and the aromatic units found in clusters **2a–2d** correspond to 1,3,5-cyclohexanetriol and 1,3,5-benzenetriol, respectively. Although in the present study only structures with etheric bonds were investigated, other linkers (i.e., C–N–C bonds) could also be considered to connect adjacent carbon rings. To preserve the high molecular symmetry, it is mandatory that the rings maintain their relative positions.

When they were part of the double-shell assembly, only four rings from the inner layer were covalently connected to the surface layer. Linkers were attached to the carbon atoms, which did not have a neighboring oxygen atom. Therefore, these rings had six connections, and they correspond to the hexavalent blue atoms from structure **1c**.

The core fragments were hollow on the inside and could encapsulate smaller molecules or metal atoms, which could enhance targeted drug delivery [17].

3.1.2. Shell Building Blocks

Figure 3 shows possible candidate molecules **3a–3d**, which correspond to the outer layer in the assembly of the double-shell structures. Their energy-minimized geometry (Figure 3) shows that they have a spherical shape and are hollow molecules that can accommodate guest molecules **2a–2f**.

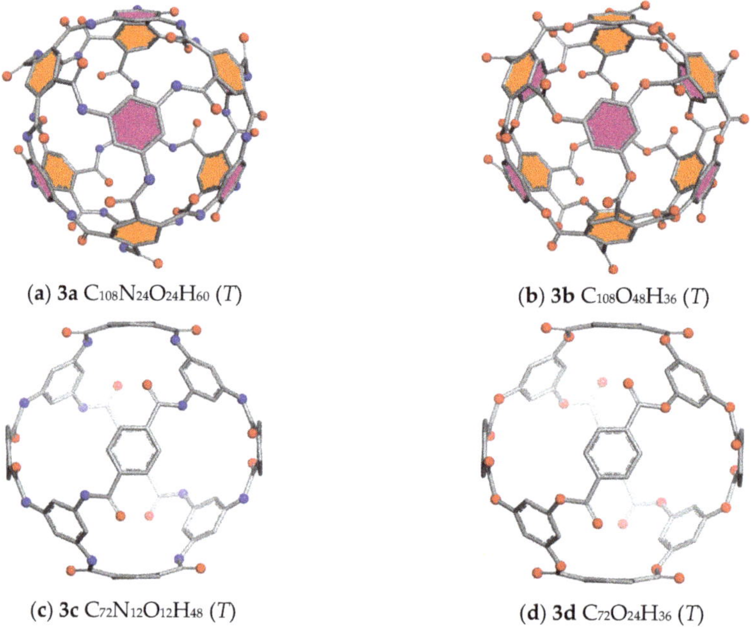

(**a**) **3a** $C_{108}N_{24}O_{24}H_{60}$ (T)   (**b**) **3b** $C_{108}O_{48}H_{36}$ (T)

(**c**) **3c** $C_{72}N_{12}O_{12}H_{48}$ (T)   (**d**) **3d** $C_{72}O_{24}H_{36}$ (T)

Figure 3. Optimized geometries of the surface layers. The two types (distinct number of connections) of aromatic rings are highlighted in violet and orange, respectively. Molecules **3a** and **3b** are viewed along the C_3 rotation axis, whereas **3c** and **3d** are aligned along the twofold rotation axis.

Only aromatic rings were included in the carbon framework and were connected by means of amide (**3a** and **3c**) or ester (**3b** and **3d**) bonds. There were two symmetry distinct rings, which were connected in an alternating pattern. One type of benzene ring had three neighbors and were aligned along the threefold rotation axis (marked in violet in Figure 3), whereas those with four adjacent aromatic rings were positioned along the twofold symmetry axis (highlighted in orange in Figure 3). Structures **3a** and **3b** were composed of 14 aromatic rings; the removal of four three-connected rings resulted in their corresponding molecules **3c** and **3d**, respectively.

Covalent connection with the core layer was realized by the attachment of functional groups to the two non-substituted carbon atoms in the orange benzene rings. These aromatic rings correspond to the hexavalent red vertices from cube–rhombellane **1a**; accordingly, each carbon was a junction point to a neighboring ring.

3.1.3. Double-Shell Assemblies

The covalent clusters shown in Figure 4 were designed by the connection of the corresponding core and shell fragments by means of –CH_2O– linkers. Twelve junctions linked the two building blocks by means of covalent bonds. The studied assemblies are shown in Figure 4, where the carbon framework of the inner layers is highlighted in green. Clusters **4a** and **4b** were achieved by joining together a core fragment **2f** with shell structures **3c** and **3d**, respectively. Since molecule **2f** included both aromatic and aliphatic rings, the linkers were attached only to the cyclohexane rings. Energetically favorable assembly between the core layer **2c** and shell fragment **3a** resulted in structure **4c**.

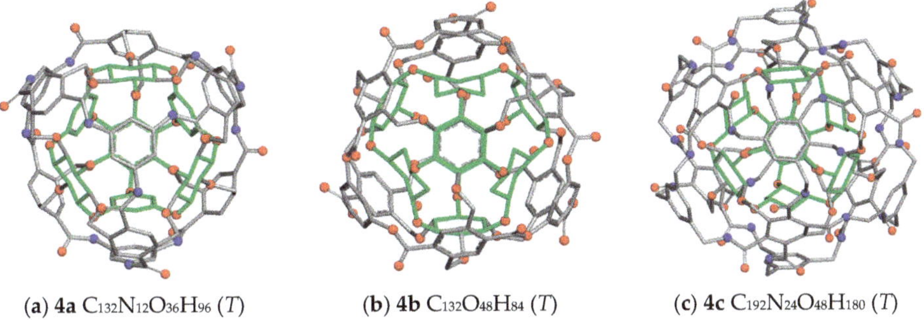

(**a**) **4a** $C_{132}N_{12}O_{36}H_{96}$ (T)   (**b**) **4b** $C_{132}O_{48}H_{84}$ (T)   (**c**) **4c** $C_{192}N_{24}O_{48}H_{180}$ (T)

Figure 4. Optimized geometries of double-shell cube–rhombellane structures where the core and shell building blocks are connected by –CH_2O– linkers. In each structure, the core fragment atoms are highlighted in green.

3.2. Computational Results

Each structure was energy minimized using the B3LYP functional and the 6-31G(d,p) basis set. The obtained computational results, the HOMO–LUMO energy gaps (E_{gap} in eV) and heat of formations (H_f in a.u.), are collected in Table 1. The heat of formation (binding energy) was evaluated as the difference between the energy of the molecule and its constituent atoms.

Comparison of the energies of structures **2a**–**2d** with the same chemical formula indicates that structure **2c** is the most thermodynamically (H_f = −25.14 a.u.) stable isomer. Among the cycloaliphatic isomers, **2a** has the highest heat of formation (−24.9 a.u.) and also the lowest energy gap (6.14 eV), which could be explained by the repulsion between the axial hydrogen atoms connected to the carbon atoms that are part of the ether bonds. Structure **2b**, where all oxygen atoms were connected in axial positions, has the highest energy gap (7.26 eV).

Aromatic clusters **2e** and **2f** were energetically less favorable, which could be associated with the strain of the aromatic carbon rings.

The shell molecules **3a**–**3d** had smaller energy gaps, and the heat of formation per heavy atom was also higher. Between the amidic and esteric clusters, the former showed an improved stability. The lowest energy structure was **3c** with a heat of formation per atom of −223.48 kcal/mol.

Table 1. Symmetries, HOMO–LUMO energy gaps (E_{gap} in eV), heat of formation (H_f in a.u.), and heat of formation divided by the number of heavy atoms (H_f/N in kcal/mol), obtained at the B3LYP/6-31G(d,p) level of theory.

Structure	Formula	N_{atoms}	Symm.	E_{gap} (eV)	H_f (a.u.)	H_f/N (kcal/mol)
2a	$C_{48}O_{12}H_{72}$	132	O_h	6.139	−4.906	−260.480
2b	$C_{48}O_{12}H_{72}$	132	O_h	7.264	−25.000	−261.466
2c	$C_{48}O_{12}H_{72}$	132	T_d	6.448	−25.142	−262.948
2d	$C_{48}O_{12}H_{72}$	132	O_h	6.843	−25.041	−261.895
2e	$C_{48}O_{12}H_{24}$	84	O_h	5.899	−20.120	−210.428
2f	$C_{48}O_{12}H_{48}$	108	T_d	6.200	−22.601	−236.371
3a	$C_{108}N_{24}O_{24}H_{60}$	216	T	2.877	−52.610	−211.622
3b	$C_{108}O_{48}H_{36}$	192	T	3.914	−47.180	−189.782
3c	$C_{72}N_{12}O_{12}H_{48}$	144	T	3.880	−34.190	−223.483
3d	$C_{72}O_{24}H_{36}$	132	T	4.201	−31.507	−205.949
4a	$C_{132}N_{12}O_{36}H_{96}$	276	T	3.942	−62.110	−216.525
4b	$C_{132}O_{48}H_{84}$	264	T	4.020	−59.334	−206.849
4c	$C_{192}N_{24}O_{48}H_{180}$	444	T	4.738	−95.962	−228.095

All covalent assemblies have tetrahedral (T) point group symmetries. In clusters **4a** and **4b**, the aromatic rings were located above each other, with an inter-shell spacing between the carbon atoms of 3.16 Å and 3.08 Å in case of clusters **4a** and **4b**, respectively. The short distance suggests that a stacking interaction exists between the two shells, the attractive van der Waals interactions between the benzene rings has a contribution to the energy of the clusters. Notice that, in the case of graphite, the observed spacing between two layers is 3.34–3.5 Å, whereas in the case of fullerene C_{20} encapsulated inside of C_{60}, the computed inter-shell distance is only 1.95 Å [18].

Among the double-shell clusters, structure **4c** was found to be the most stable assembly with the highest energy gap (4.74 eV) and lowest energy (H_f/N = −228.09 kcal/mol). Comparing clusters **4a** and **4b** with an identical number of heavy atoms, the amidic structure has a 10 kcal/mol energy gain with respect to the molecule containing ester bonds. The HOMO–LUMO gaps of both molecules are very close to 4 eV.

Although several other rhombellanes were built, due to geometric strain they were not energetically feasible structures. As previously mentioned, stability of the double-shell cluster is related to the optimal size of both layers, and the proper orientation of the covalent connection points. Obviously, a larger surface layer could accommodate more easily the core fragments, however longer chemical linkers are required to covalently connect the two layers.

4. Conclusions

Using the rhombellane framework, several dual-layer covalent assemblies were designed as potential drug delivery systems. The aromatic moieties through stacking interactions, as well as the hydrogen bond donors and acceptors groups on the surface layer, significantly contribute to the ligand binding capacity. The immobilization of compounds with pharmaceutical potential could be further enhanced by attachment of functional groups to the aromatic rings.

The geometry and stability of the building blocks and some covalent assemblies were investigated by means of density functional theory, using the B3LYP functional and the 6-31G(d,p) basis set. The results show that all structures are energetically feasible and the high molecular symmetry is preserved during geometry optimization. Two important geometric aspects were observed when building models of rhombellanes. First, the surface layer should be large enough to easily accommodate the core fragment, and second, the key atoms which the linkers bind covalently should be properly oriented. Otherwise, due to geometric strain, the assembly is not energetically favorable.

Assembly of the building blocks is favored by stacking interactions between the aromatic rings. However, due to their level of complexity, the chemical synthesis remains a great challenge.

Symmetry **2020**, *12*, 343

Supplementary Materials: The following are available online at http://www.mdpi.com/2073-8994/12/3/343/s1: geometries (Cartesian coordinates) of structures **2a–2f**, **3a–3d**, and **4a–4c**, optimized at the B3LYP/6-31G(d,p) level of theory.

Author Contributions: Conceptualization, C.L.N.; Methodology, C.L.N.; Investigation, K.N. and C.L.N.; Writing—Original Draft Preparation, C.L.N.; Writing—Review & Editing, C.L.N. and B.S.; Visualization, K.N.; Supervision, C.L.N.; Funding Acquisition, C.L.N. and B.S. All authors have read and agreed to the published version of the manuscript.

Funding: This research was funded by GEMNS project granted in the European Union's Seventh Framework Program under frame of the ERA-NET EuroNanoMed II (European Innovative Research and Technological Development Projects in Nanomedicine). Funding number PN-III-ID-57, no. 8/2015.

Conflicts of Interest: The authors declare no conflict of interest.

References

1. Canceill, J.; Lacombe, L.; Collet, A. A new cryptophane forming unusually stable inclusion complexes with neutral guests in a lipophilic solvent. *J. Am. Chem. Soc.* **1986**, *108*, 4230–4232. [CrossRef]

2. Cram, D.J. Molecular container compounds. *Nature* **1992**, *356*, 29–36. [CrossRef]

3. Sun, Q.F.; Murase, T.; Sato, S.; Fujita, M. A sphere-in-sphere complex by orthogonal self-assembly. *Angew. Chem. Int. Ed.* **2011**, *50*, 10318–10321. [CrossRef] [PubMed]

4. MacGillivray, L.R.; Atwood, J.L. A chiral spherical molecular assembly held together by 60 hydrogen bonds. *Nature* **1997**, *389*, 469–472. [CrossRef]

5. Liu, Y.; Hu, C.; Comotti, A.; Ward, M.D. Supramolecular archimedean cages assembled with 72 hydrogen bonds. *Science* **2011**, *333*, 436–440. [CrossRef] [PubMed]

6. Grimme, S.; Mück-Lichtenfeld, C.; Antony, J. Noncovalent interactions between graphene sheets and in multishell (hyper)fullerenes. *J. Phys. Chem. C* **2007**, *111*, 11199–11207. [CrossRef]

7. Casella, G.; Bagno, A.; Saielli, G. Spectroscopic signatures of the carbon buckyonions $C_{60}@C_{180}$ and $C_{60}@C_{240}$: A dispersion–corrected DFT study. *Phys. Chem. Chem. Phys.* **2013**, *15*, 18030–18038. [CrossRef] [PubMed]

8. Pichierri, F. Hypercubane: DFT-based prediction of an O_h-symmetric double-shell hydrocarbon. *Chem. Phys. Lett.* **2014**, *612*, 198–202. [CrossRef]

9. Pichierri, F. Substituent effects in cubane and hypercubane: A DFT and QTAIM study. *Theor. Chem. Acc.* **2017**, *136*, 114. [CrossRef]

10. Diudea, M.V. Rhombellanic crystals and quasicrystals. *Iran. J. Math. Chem.* **2018**, *9*, 167–178.

11. Diudea, M.V.; Nagy, C.L. Rhombellane space filling. *J. Math. Chem.* **2019**, *57*, 473–483. [CrossRef]

12. Diudea, M.V.; Lungu, C.N.; Nagy, C.L. Cube-Rhombellane related structures: A drug perspective. *Molecules* **2018**, *23*, 2533. [CrossRef] [PubMed]

13. Czeleń, P.; Szefler, B. The Immobilization of ChEMBL474807 Molecules Using Different Classes of Nanostructures. *Symmetry* **2019**, *11*, 980. [CrossRef]

14. Czeleń, P.; Szefler, B. The immobilization of oxindole derivatives with use of cube rhombellane homeomorphs. *Symmetry* **2019**, *11*, 900. [CrossRef]

15. Szefler, B.; Czeleń, P. Docking of cisplatin on fullerene derivatives and some cube rhombellane functionalized homeomorphs. *Symmetry* **2019**, *11*, 874. [CrossRef]

16. Frisch, M.J.; Trucks, G.W.; Schlegel, H.B.; Scuseria, G.E.; Robb, M.A.; Cheeseman, J.R.; Scalmani, G.; Barone, V.; Mennucci, B.; Petersson, G.A.; et al. *Gaussian 09, Revision E.01*; Gaussian, Inc.: Wallingford, CT, USA, 2009.

17. Liu, Y.-L.; Chen, D.; Shang, P.; Yin, D.-C. A review of magnet systems for targeted drug delivery. *J. Control. Release* **2019**, *302*, 90–104. [CrossRef] [PubMed]

18. Liu, F.; Meng, L.; Zheng, S. Density functional studies on a novel double–shell fullerene $C_{20}@C_{60}$. *J. Mol. Struct. THEOCHEM* **2005**, *725*, 17–21. [CrossRef]

Article

The Immobilization of ChEMBL474807 Molecules Using Different Classes of Nanostructures

Przemysław Czeleń * and Beata Szefler

Department of Physical Chemistry, Faculty of Pharmacy, Collegium Medicum, Nicolaus Copernicus University, Kurpinskiego 5, 85-096 Bydgoszcz, Poland
* Correspondence: przemekcz@cm.umk.pl

Received: 24 June 2019; Accepted: 23 July 2019; Published: 2 August 2019

Abstract: Indirubin derivatives and analogues are a large group of compounds which are widely and successfully used in treatment of many cancer diseases. In particular, the ChEMBL474807 molecule, which has confirmed inhibiting abilities against CDK2 and GSK3B enzymes, can be included in this group. The immobilization of inhibitors with the use of nanocarriers is an often used strategy in creation of targeted therapies. Evaluations were made of the possibility of immobilizing ligand molecules on different types of nanocarrier, such as carbon nanotubes (CNT), functionalized fullerene C_{60} derivatives (FF_X), and functionalized cube rhombellanes, via the use of docking methods. All results were compared with a reference system, namely C_{60} fullerene. The realized calculations allowed indication of a group of compounds that exhibited significant binding affinity relative to the ligand molecule. Obtained data shows that structural modifications, such as those related to the addition of functional groups or changes of structure symmetry, realized in particular types of considered nanostructures, can contribute to increases of their binding capabilities. The analysis of all obtained nano complexes clearly shows that the dominant role in stabilization of such systems is played by stacking and hydrophobic interactions. The realized research allowed identification of potential nanostructures that, together with the ChEMBL474807 molecule, enable the creation of targeted therapy.

Keywords: indirubin analogs; docking; rhombellanes; fullerenes; ChEMBL474807; C_{60}; carbon nanotubes

1. Introduction

Indirubin is a natural compound which can be found as a component of natural indigo, and for many years has been well known as an ingredient used in many traditional medicine methods. Its applications are numerous and related to its pharmacological potential, manifesting through anti-inflammatory [1], antitumor [2,3], antiangiogenic [4], and neuroprotective [5] effects. Indirubin and its derivatives or analogues are also well known as competitive ATP inhibitors; such inhibiting potential has also been confirmed in the case of cyclin-dependent kinases (CDK) and glycogen synthase kinase 3 (GSK3) [6,7], and contributed to the development of promising drugs for diabetes, inflammation, cancer, and neurodegeneration [5,8–11]. The ChEMBL474807 molecule can be included in this group of compounds, since its inhibitory properties against CDK2 and GSK3B have been confirmed with the use of molecular modeling methods [12]. The direct application of compounds with pharmacological potential can be fraught, with many potential side effects related, for example, to toxicity or wide spectrum of impact. One of the most commonly used strategies for prevention of such phenomena is the preparation of targeted therapy involving the use of carriers providing controlled release of drugs. Such therapies not only increase the efficiency and bioavailability of the drugs used, but also often reduce the occurrence of side effects [13–17]. The chemical structure of the ChEMBL474807 ligand molecule is presented in Figure 1. Four cyclic systems containing three aromatic rings indicate that one of the most important stabilizing effects in interactions with potential carriers should be stacking or hydrophobic interactions. The presence of four hydrogens which could play the role of hydrogen bond donors also is worth taking into account. A large group of commonly used carriers are compounds included in the group of fullerenes or their derivatives. One of the most intensively used fullerene molecules is C_{60}. Numerous studies have shown significant binding potential of this particle in the context of π–π stacking interactions with many biologically active compounds [18]. The chemical structure of such a molecule also enables free penetration through cell membranes [17,19,20], however, small concentrations of this compound are non-toxic for living organisms [13,21–23]. The structure of the C_{60} molecule exhibits structural disproportion compared to the considered ligand molecule, and the lack of any hydrogen bond acceptors also encourages the search for new nanostructures that could better utilize the binding potential of the inhibitor.

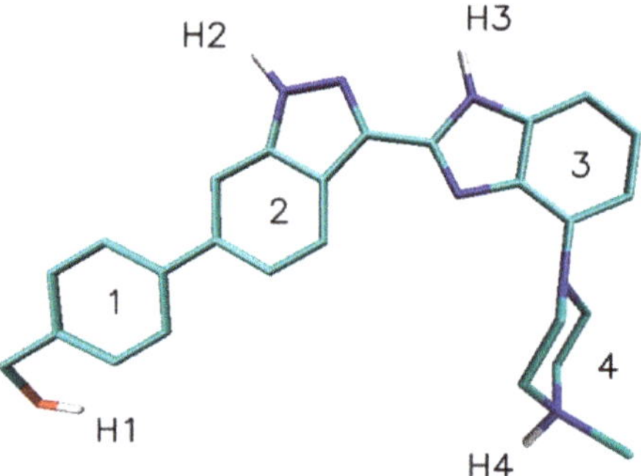

Figure 1. Graphic representation of ligand molecule.

The most intuitive solution seems to be analysis of the functionalized derivatives of C_{60} fullerene, which have found numerous applications in modern medicine [24–26]. Table S1 presents sets of chemical structures and names of proposed fullerenes. In this group of nanocarriers, compounds with single and multiple substituents placed on the surface of C60 fullerene can be found. The scope of the functional groups used takes into account aliphatic chains with methoxy, amide, ester, and halogen groups, and aromatic systems that include unsubstituted benzene rings, as well as more complex aromatic systems. The next set of nanosystems worthy of consideration is carbon nanotubes, which have found many applications in the biomedical field e.g., in diagnostics, tissue engineering, or targeted drug delivery [27–29]. The characteristic structure of these nanosystems is created by rolling up graphene sheets, and such large surfaces consisting of aromatic systems could exhibit high affinity toward the considered ligand molecule that exhibiting significant activity in stacking interactions. The last interesting group of nanocarriers are the cube rhombellane (Cube-rbl) homeomorphs, consisting of a molecule core and a second shell created by eight modules connected with the structure by three chemical bonds, e.g., ester or amide [30–33]. The examples of such structures created by Diudea are presented in Figure 2. The proposed nanostructures, due to their characteristic modular structures, provide numerous active groups arranged on a nanocarrier surface in an organized manner, strictly dependent on the symmetry of the molecule core. Depending on the type of the particular molecule, the active groups are aromatic rings or cyclohexane derivatives and ester or amide groups. The precise characteristics of all Cube-rbl are displayed in Table S2. The evaluation of ligand molecule affinity relative to all proposed nanocarriers can contribute to the creation of effective targeted therapy.

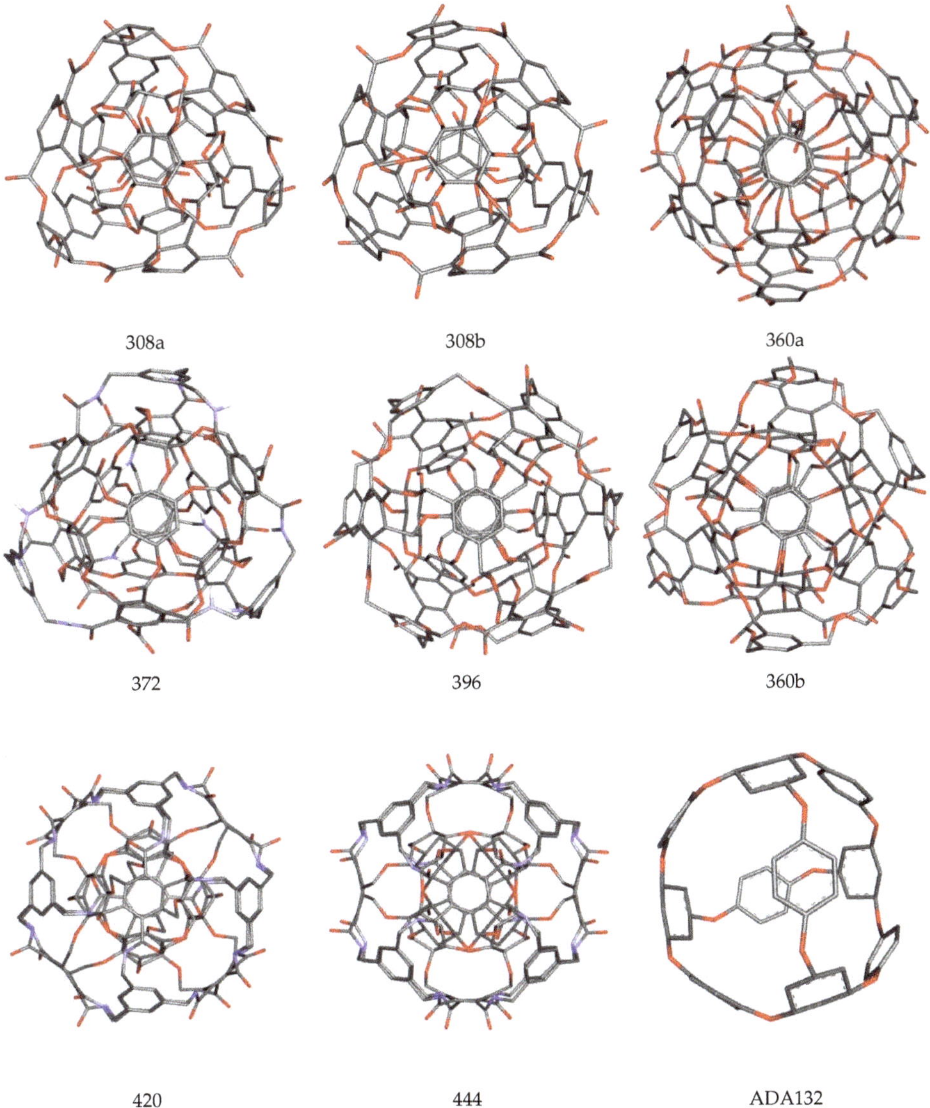

Figure 2. Graphic representation of the chosen structures of cube rhombellane homeomorphs.

2. Methods

The ligand molecule, namely ChEMBL474807 (1-(4-amino-1,2,5-oxadiazol-3-yl)-5-(piperidin-1-ylmethyl)-*N*′-(pyridin-4-ylmethylene)-1H-1,2,3-triazole-4-carbohydrazide), was obtained from CHEMBL Database [34]. Rhombellane homeomorphs, were obtained from Topo Cluj Group [33], the C_{60} structure was downloaded from Brookhaven Protein Database PDB [35], and C_{60} functionalized derivatives were obtained from the PubChem database [36]. The carbon nanotubes described in Table 1 were generated using the VMD package [37], chosen values of "n" and "m" vectors allowed for the creation of different types of nanotubes characterized by similar diameter values. The AutoDockVina program and implemented united-atom scoring function was used during the docking stage [38].

Table 1. Description of carbon nanotube parameters used during docking stage.

Nanostructure Name	C Atoms	N chiral Index	M Chiral Index	L [nm]	Type	D [Å]	Diameter Class
CNT_260	260	5	5	3	armchair	6.25	D1
CNT_252	252	6	3	3	chiral	6.21	D1
CNT_256	256	8	0	3	zigzag	6.73	D1
CNT_416	416	8	8	3	armchair	10.83	D2
CNT_392	392	10	6	2	chiral	10.94	D2
CNT_448	448	14	0	3	zigzag	10.94	D2
CNT_676	676	13	13	3	armchair	17.47	D3
CNT_760	760	15	10	3	chiral	17.01	D3
CNT_704	704	22	0	3	zigzag	17.20	D3

During the docking stage, ligand and nanosystem structures containing only polar hydrogen atoms were used. In the case of all "spherical" nanoparticles, the grid box dimensions were established equal to $26 \times 26 \times 26$ Å, while in the case of carbon nanotubes, sets of values were chosen that ensured free interactions of the ligand molecule with the surface and interior of the nanosystems used. All initial procedures related to preparation of the ligand and nanosystems during the docking procedure were realized with the use of AutoDock Tools package [39]. All calculations during the docking stage were realized with an exhaustiveness parameter equal to 20, since such a value ensures an appropriate reproducibility of the results and a reasonable time of calculation. Structural analysis of the considered systems and visualization of the obtained complexes were realized with use of the VMD package [37]. The value of the binding constant was calculated based on the formula:

$$K_{max} = exp^{\left(\frac{-\Delta G_{max}}{RT}\right)} \tag{1}$$

where ΔG_{max} represents maximal value of binding affinity obtained during docking stage, R represents value of gas constant, and T represents temperature.

3. Results and Discussion

The affinity of a ligand molecule towards different classes of nanocarrier was evaluated with the use of docking methods. The carbon nanoparticle most often used as a nanocarrier is the C_{60} molecule, therefore, it was used as a reference system in this study. The binding affinity of the considered ligand molecule towards native C_{60} molecules was equal to -4.90 kcal/mol (Table 2). The structure of the obtained complex presented in Figure 3 shows that the long structure of the ligand molecule has limited ability to interact with fullerene C_{60} and cannot effectively use all aromatic systems in creation of interactions. The observed single stacking interaction of one of the aromatic systems of the ligand molecule cannot strongly stabilize the analyzed complex. Table 2 also presents the values obtained for functionalized C_{60} fullerenes. Structural changes observed in the considered molecules were related to the addition of various functional groups, including aliphatic chains with methoxy, amide, ester, and halogen groups supporting the possibility of creation of hydrogen bonds and aromatic systems that include unsubstituted benzene rings, as well as more complex aromatic systems (Table S1). All considered compounds exhibited much higher binding affinity towards ligand molecule than native C_{60} fullerene, which manifested in the increase of these values by -1.0 to -3.1 kcal/mol. The differences in binding affinity were more evident in the context of the binding constant of complex formation [K_{max}], for which the observed increase relative to the reference system started from 440% and in the best cases even reached 18,621% (Table 2). The most interesting outcomes were obtained for functionalized fullerenes containing aromatic substituents. The presence of a single aromatic group increased the values of binding affinity by at least -1.7 kcal/mol. The possibility of additional stabilizing impacts significantly affected the durability of considered complexes. Graphic representations of the most interesting complexes are shown in Figure 3. The presence of a single aromatic functional group on the surface of functionalized C_{60} molecule, observed for example in the case of FF_5, FF_7, FF_8, and FF_11 molecules, forced a characteristic orientation of the ligand molecule. The first aromatic system of the ligand molecule interacted with aromatic systems from functional groups of the considered fullerenes, and the observed planar orientations, with distances in the range of 3.63 to 3.68 Å, indicated strong stacking interactions. The next stabilizing impact was related to the second aromatic system of the ligand molecule, the orientation of which enabled interactions with the aromatic surface of the fullerene core, and, in this case, the distances between aromatic systems were smaller and fell in the range of 3.48 to 3.57 Å. The conformation of ligand molecules in the considered complexes caused the third aromatic system to be unable to actively interact with the surface of the fullerenes. The last stabilizing impact was related to the fourth cyclic system of the ligand molecule, of which the orientation enabled the appearance of hydrophobic interactions with the fullerene surface. The addition of multiple aromatic systems on the fullerene surface, like in the case of FF_12, significantly increased the stability of the obtained complexes; this was confirmed by the values of the binding constant [K_{max}], which reached values 18,621% higher than the reference system. The presence of multiple aromatic substituents on the fullerene surface of the FF_12 molecule allowed all aromatic systems of the ligand molecule to be engaged in the creation of stacking interactions (Figure 3). Many functionalized fullerenes used during the docking stage contained hydrogen bond acceptors, however, the creation of such interactions with any polar hydrogen atom of the ligand molecule was not observed in any case. Such an observation clearly indicates the dominant role of stacking and hydrophobic interactions in stabilization of ligand complexes with functionalized fullerenes.

Table 2. Values of binding affinity [kcal/mol] of ligand molecule relative to functionalized C_{60} fullerene derivatives (FF_X) obtained during docking stage.

Nanostructure Name	ΔG [kcal/mol]				Binding Constant [K_{max}]	Difference of K_{max} Relative to C_{60} [%]
	MAX	**MIN**	**AVERAGE**	**SD**		
FF_1	−6.50	−6.00	−6.24	0.13	58,151.8	1388.7
FF_2	−6.10	−5.80	−5.96	0.09	29,604.6	657.9
FF_3	−6.50	−6.10	−6.31	0.11	58,151.8	1388.7
FF_4	−6.00	−5.90	−5.95	0.05	25,006.8	540.2
FF_5	−7.10	−6.50	−6.74	0.16	160,091.8	3998.5
FF_6	−6.10	−5.70	−5.85	0.13	29,604.6	657.9
FF_7	−6.80	−6.40	−6.57	0.10	96,486.4	2370.1
FF_8	−6.70	−6.30	−6.50	0.12	81,501.4	1986.5
FF_9	−6.60	−6.40	−6.48	0.07	68,843.7	1662.4
FF_10	−6.10	−5.70	−5.82	0.11	29,604.6	657.9
FF_11	−7.20	−6.70	−6.92	0.15	189,526.5	4752.0
FF_12	−8.00	−7.40	−7.66	0.18	731,270.0	18,621.0
FF_13	−5.90	−5.80	−5.83	0.05	21,123.1	440.8
C_{60}	−4.90	−4.90	−4.90	0.00	3906.1	0.0

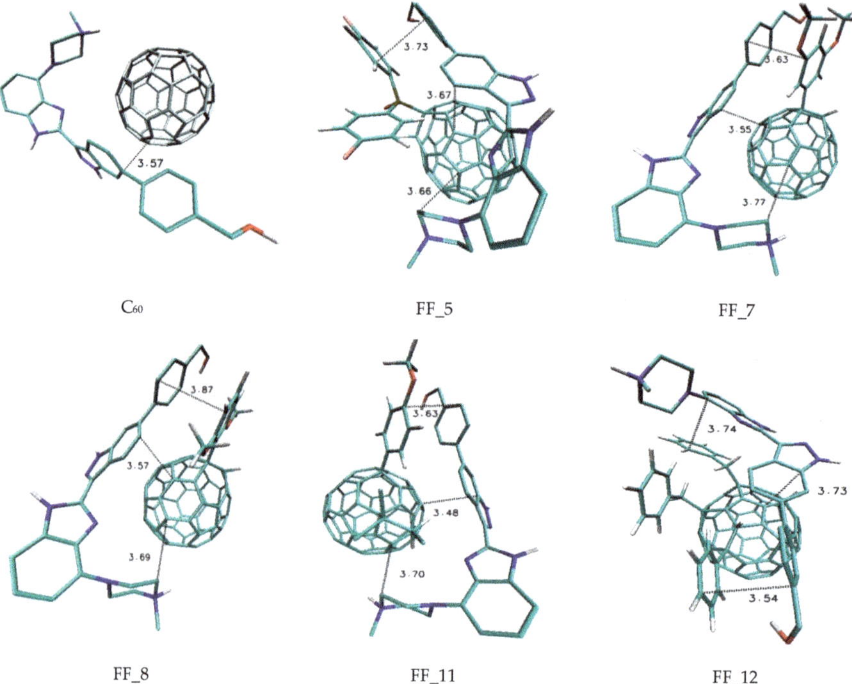

Figure 3. Graphic representation of ligand complexes with functionalized derivatives of C_{60} fullerene characterized by highest binding affinity.

The next class of nanostructures used during modeling were carbon nanotubes. Three types of such nanosystems were used, namely zigzag (n, 0), armchair (n, n), and chiral (n, m), with each set of vector parameters defining the way the graphene sheet is rolled. Appropriate graphic representations for each type of nanotube are shown in Figure 4. The values presented in Table 3 describe the affinity of the ligand molecule for each type of nanotube considered, characterized by three different diameter values. For each nanotube, the ligand affinity relative to the surface and interior of the considered nanosystem was evaluated. In the context of interaction with the surface of nanotubes, one of the most important factors is the nanotube diameter, and presented data clearly show that three classes of systems can be distinguished.

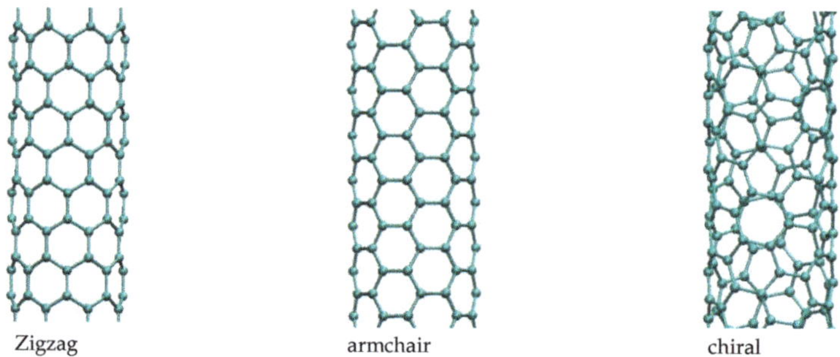

Figure 4. Graphic representation of nanotube structure types.

Table 3. Values of binding affinity [kcal/mol] of the ligand molecule relative to carbon nanotubes (CNT) obtained during docking stage. The values presented in *italics* are positive and indicate the non-physical character of the interactions of the ligand molecule with the interior of the nanotube with smallest diameter.

Nanostructure Name	ΔG [kcal/mol]				Binding Constant [K_{max}]	Type
	MAX	**MIN**	**AVERAGE**	**SD**		
CNT surface						
CNT_260	−10.8	−10.6	−10.7	0.06	8.25×10^7	armchair
CNT_252	−10.5	−10.3	−10.4	0.06	4.97×10^7	chiral
CNT_256	−10.4	−10.3	−10.4	0.05	4.20×10^7	zigzag
CNT_416	−11.6	−11.4	−11.5	0.06	3.18×10^8	armchair
CNT_392	−11.7	−11.5	−11.6	0.07	3.77×10^8	chiral
CNT_448	−11.6	−11.5	−11.6	0.05	3.18×10^8	zigzag
CNT_676	−12.6	−12.4	−12.5	0.05	1.72×10^9	armchair
CNT_760	−12.6	−12.4	−12.5	0.05	1.72×10^9	chiral
CNT_704	−12.5	−12.4	−12.4	0.05	1.45×10^9	zigzag
CNT interior						
CNT_260	*134.9*	*142*	*138*	*2.6*	1.31×10^{-99}	*armchair*
CNT_252	*249.9*	*269*	*257*	*10.9*	6.62×10^{-184}	*chiral*
CNT_256	*232*	*271*	*257*	*21.7*	8.74×10^{-171}	*zigzag*
CNT_416	−19	−18.9	−18.9	0.02	8.46×10^{13}	armchair
CNT_392	−19.4	−19.2	−19.3	0.09	1.66×10^{14}	chiral
CNT_448	−19.1	−19	−19	0.05	1.00×10^{14}	zigzag
CNT_676	−18.3	−18.2	−18.2	0.04	2.59×10^{13}	armchair
CNT_760	−18.5	−18.3	−18.4	0.06	3.64×10^{13}	chiral
CNT_704	−18.4	−18.3	−18.4	0.04	3.07×10^{13}	zigzag

Increasing nanotube diameter and the related reduction of their surface curvature clearly affected the affinity relative to the ligand molecule. For each change of nanotube diameter from ~6Å to ~11 Å and to ~17 Å, simultaneous increases of binding affinity by about −1 kcal/mol were observed. In the case of two classes of nanotubes characterized by higher diameter values, the type of nanotube did not influence the affinity toward the ligand. Another situation was observed for nanotubes characterized by the smallest diameter values, as in this case, differences in affinities of particular types of nanotubes occured. Relatively large nanotube surface curvatures preferred the "armchair" type, which exhibited the highest affinity toward the ligand molecule. Differences of about −0.4 kcal/mol relative to other types of nanotubes indicated that the binding constant for that complex was twofold higher. The structures of the ligand molecule complexes with the chosen nanotubes are presented in Figure 5. In the case of systems where the ligand molecule interacted with the surface of the nanotube, a longitudinal orientation of considered molecules was always observed, regardless of the diameter of the nanotube. This orientation was dominant for all considered complexes. The planar orientation of the ligand molecule in all complexes and distances placed in the range from 3.43 to 3.74 Å caused all aromatic systems of the ligand molecule to be involved in creation of stable stacking interactions with the surfaces of the nanotubes used. The next analyzed aspect is related to interaction of the ligand molecule with nanotube interior. In this case, only two classes of nanotubes supported a sufficient diameter for this goal (~11 Å, ~17 Å). The data presented in Table 3 show that nanosystems with a smaller diameter (D2) interacted with the ligand molecule much more strongly, and the differences in binding affinities between the compared systems were in the range of −0.7 to −0.9 kcal/mol. The structures of complexes created by two compared classes of nanotubes were also significantly different (Figure 5). In the case of class D3, linearly arranged ligand molecule interacted with an internal nanotube surface, while in the case of class D2 nanotubes, centrally located ligand molecules could interact with the upper and lower plane of the molecule with nanotube walls. Additionally, the presence of aliphatic cyclic systems did not allow for an energetically favorable conformation to be obtained completely localized inside the nanotube. This part of the ligand interacted with the edge of the nanotube.

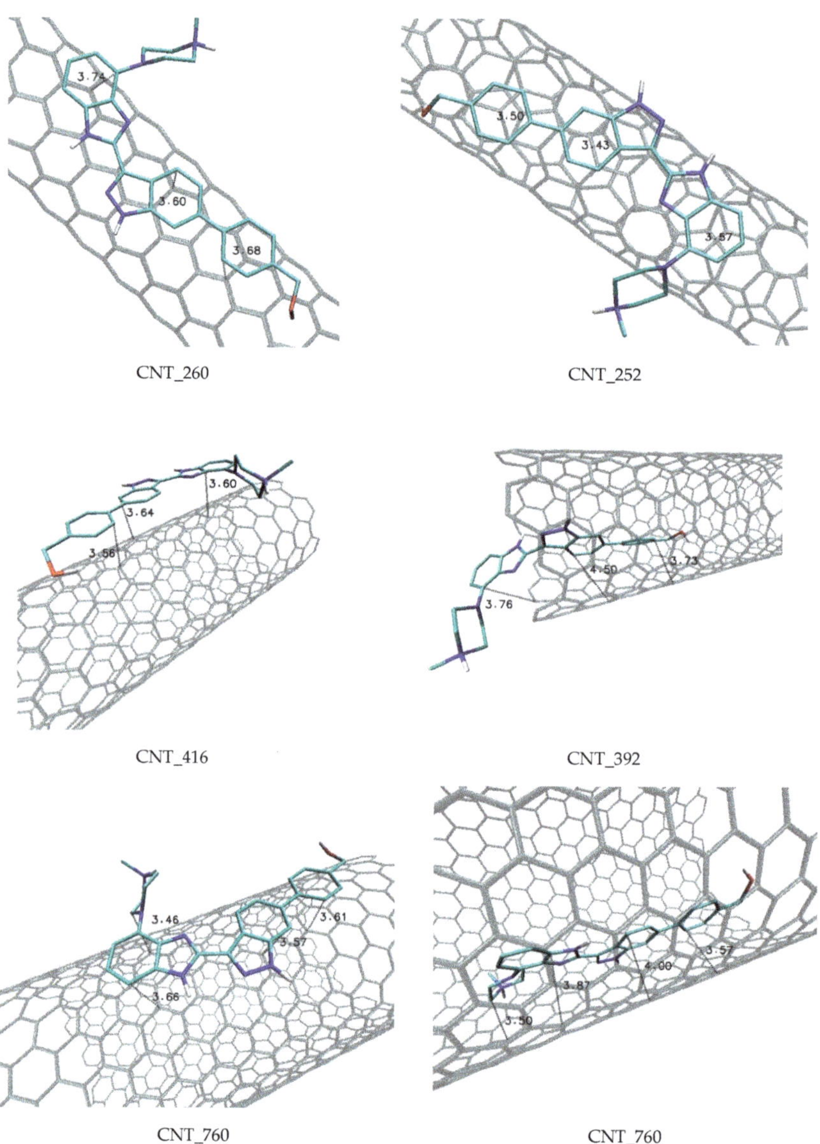

CNT_260   CNT_252

CNT_416   CNT_392

CNT_760   CNT_760

Figure 5. Graphic representation of complexes of the ligand molecule with chosen carbon nanotubes.

The last group of nanostructures utilized for the immobilization of the inhibitor molecule were cube rhombellanes and their homeomorphs. Values presented in Table 4 show that a significant group of considered nanocarriers exhibited a higher binding affinity towards the ligand molecule than the reference molecule C_{60}, and the observed differences were placed in the range of -0.7 to -1.8 kcal/mol. Calculated values of complex binding constants unambiguously show that such complexes were created much more efficiently, and the observed differences were placed in the range from 226 to 1986%. The best binding properties relative to the ligand molecule among all cube rhombellane homeomorphs were shown by the 360a molecule. This nanomolecule consists of an aromatic core connected by ether bonds and eight external shell modules connected by ester bonds. A graphic representation of the complex created by this molecule is shown in Figure 6. The first and second aromatic system of the ligand molecule interacted by stacking interactions with one of the external modules of the nanocarrier, which was confirmed by the planar orientation of the considered aromatic systems and distance 3.56 Å. The next stabilizing impact was related to the presence of medium strength hydrogen bond created by a hydrogen atom H4 with oxygen from ester group. In this case, the occurrence of a hydrophobic interaction between the ligand fourth cyclic system and an aromatic system of the nanomolecule was possible. Much lower efficiency in complex creation was shown by the second structural isomer of the 360 molecule, namely form "b." The inner core of this nanomolecule had a different symmetry, which affected the structure of the external shell. Such structural changes meant that the best obtained complex conformation was characterized by a binding affinity lower by 0.7 kcal/mol than that of the first isomer of the 360 molecule. The complex of the ligand and 360b molecule was maintained by a single stacking interaction created by the second aromatic system of the inhibitor molecule. There were also two hydrogen bonds observed; first of them, created by hydrogen H2, was a very weak one, as indicated by a length close to the limit of occurrence of hydrogen bonding, while the second one was a medium strength interaction created by hydrogen H4.

Table 4. Values of binding affinity [kcal/mol] of ligand molecule relative to spherical nanosystems obtained during docking stage.

Nanostructure Name	ΔG [kcal/mol]				Binding Constant [K_{max}]	Difference of K_{max} Relative to C_{60} [%]
	MAX	**MIN**	**AVERAGE**	**SD**		
144_ex_ex	−3.40	−3.30	−3.32	0.04	310.6	−92.0
144_in_ex	−3.80	−3.80	−3.80	0.00	610.2	−84.4
156_ex_ex	−3.60	−3.60	−3.60	0.00	435.3	−88.9
156_in_ex	−3.80	−3.80	−3.80	0.00	610.2	−84.4
308a4	−6.60	−6.40	−6.50	0.09	68,843.7	1662.4
308b4	−6.50	−6.40	−6.41	0.03	58,151.8	1388.7
360a	−6.70	−6.30	−6.49	0.15	81,501.4	1986.5
360b	−6.00	−5.90	−5.98	0.04	25,006.8	540.2
372AB	−6.10	−6.00	−6.09	0.03	29,604.6	657.9
396	−6.10	−5.90	−6.01	0.06	29,604.6	657.9
420	−5.80	−5.60	−5.70	0.05	17,842.5	356.8
444	−5.60	−5.50	−5.54	0.05	12,730.8	225.9
456	−5.60	−5.40	−5.50	0.07	12,730.8	225.9
ADA_132	−5.80	−5.70	−5.72	0.04	17,842.5	356.8
C_{60}	−4.90	−4.90	−4.90	0.00	3906.1	0.0

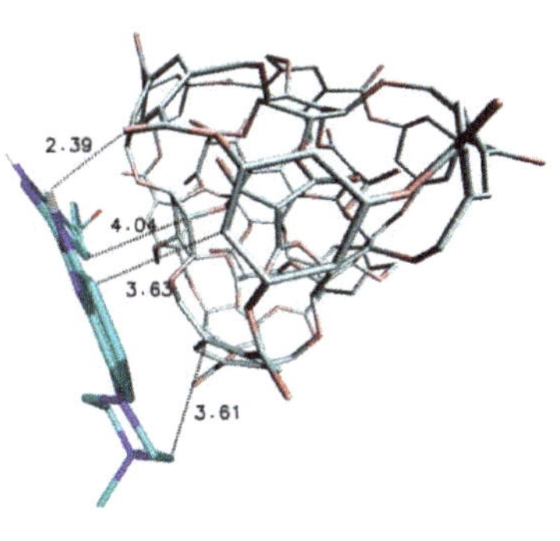

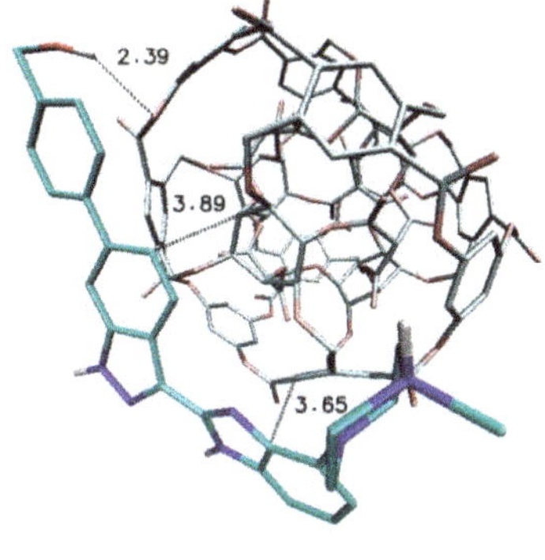

Figure 6. *Cont.*

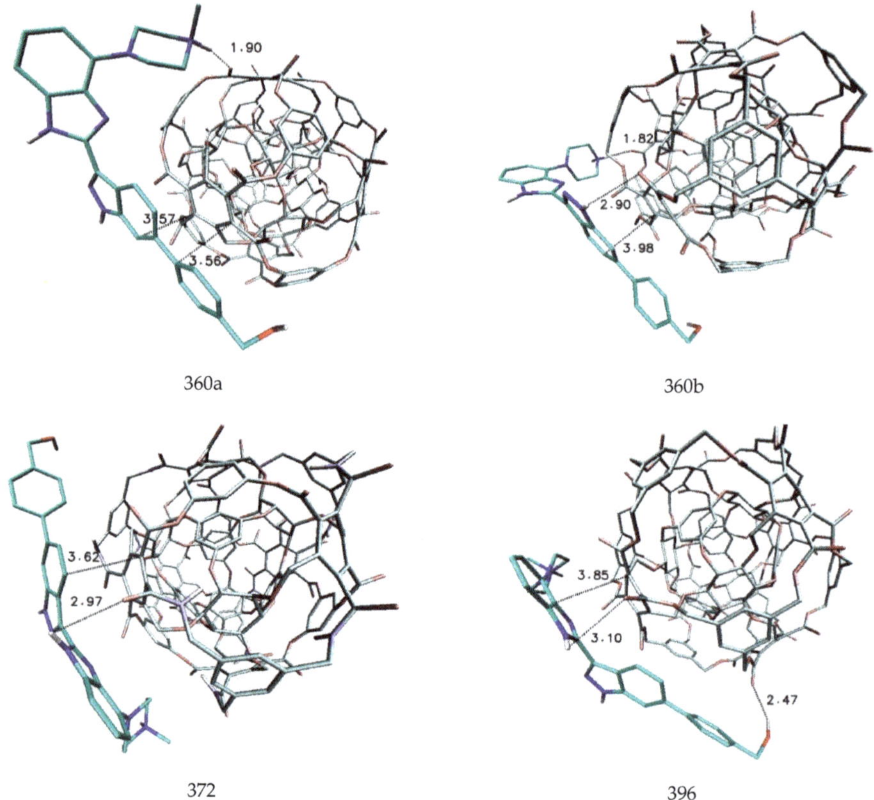

Figure 6. Graphic representation of complexes of ligand molecule with chosen cube rhombellane homeomorphs.

The next structures exhibiting significant affinity toward the considered ligand were the isomers of molecule 308 (a, b). The binding affinity for both structures reached values of at least −1.6 kcal/mol higher than for the reference molecule. In both cases, the ligand molecule was involved in stacking interactions by the second and third aromatic system. One hydrogen bond was also observed in both complexes, namely, in the case of molecule 308a, it was a weak interaction created by hydrogen H3, while another weak interaction was created by hydrogen H1 in the case of molecule 308b. All discussed interactions can be observed in Figure 6.

The last two systems that are worth discussing are molecules 372 and 396, both of which exhibited similar binding affinity towards the ligand molecule equal to −6.1 kcal/mol. The complex of the former was stabilized by only one stacking interaction, occurring between the second aromatic ring of the ligand molecule and one of aromatic modules from nanocarrier external shell, and one weak hydrogen bond which was localized between hydrogen H3 and the oxygen from the amide group of the nanomolecule. The characteristics of interactions stabilizing the complex of molecule 396 were quite similar; in this case, the third aromatic system of the ligand molecule was involved in stacking interactions, along with one weak hydrogen bond, created by hydrogen H1.

4. Conclusions

The immobilization of compounds with pharmaceutical potential serves many functions, such as decreasing their toxicity, elongating their bioavailability, and increasing their pharmacological effect. The possibility of achieving such effects is strictly related to an appropriate choice of nanocarrier to be used in creation of such nanodrug. In this work, three different classes of possible nanostructures which could be used in immobilization of the considered ligand molecule were used. The characteristic structure of the inhibitor molecule, consisting of four cyclic systems, of which three are aromatic, clearly showed that stacking and hydrophobic interactions should play a dominant role in stabilization of interactions with such a molecule. The proposed groups of compounds made it possible to unambiguously verify this thesis. The highest values of both binding affinity and binding constant were found in the case of carbon nanotubes. Such nanosystems, being a rolled up graphene plane, provided a possibility of the occurance of stacking interactions with the ligand molecule rich in aromatic systems, exhibiting the ability to obtain a planar conformation. The analysis of systems with different diameter allowed the conclusion that the increase of this parameter, with an accompanying decrease in the curvature of the nanotube surface interacting with ligand molecule, clearly contributes to increased binding capacity of such nanosystems. Significant binding capacities of such systems are dictated by the considerable size of the considered nanostructures, which may be one of the factors limiting their applicability.

The next group of studied systems were functionalized derivatives of fullerene C_{60}. The complexes of ligand molecules with such nanocarriers were characterized by much lower values of binding affinity than those created by CNT. This phenomenon can be largely explained by the size of the considered nanoparticles and the geometry of the molecules determining the exposure of the fullerene aromatic systems. The analyzed functionalized fullerenes clearly demonstrated the group of potential modifications that exhibit significant potential in increasing of binding capacity of C_{60} molecule. The addition of a single aromatic system on the surface of the fullerene considerably increased binding affinity toward the ligand molecule, relative to native fullerene. The addition of multiple aromatic systems contributed to the achievement of even better binding abilities of the considered nanosystems. Another class of modifications were substituents containing hydrogen bond donors and acceptors; however, the observed effects were less satisfactory, which was due to lower affinity values relative to nanostructures with aromatic substituents and the lack of any conformation of ligands that ensured the formation of hydrogen bonds. Such observations clearly confirm that in the case of this particular ligand molecule, the most important interactions are stacking interactions between aromatic moieties.

The last group of considered nanosytems, namely cube rhombellane homeomorphs, showed the lowest binding capacity relative to the ligand molecule. A large group of proposed compounds exhibited satisfying binding affinity towards the ligand molecule. The binding capacity of these nanostructures is strictly related with the structure of the external shell of these molecules. The example of molecules 360a and 360b clearly shows that two structural isomers can exhibit significant differences in this field.

Supplementary Materials: The following are available online at http://www.mdpi.com/2073-8994/11/8/980/s1, Table S1: Description of functionalized C_{60} fullerene derivatives (FF_X) used during docking stage, Table S2: Description of cube rhombellanes (Core) and functionalized structures with external layer (C-rbl). Value chem. type define chemical bond type characteristic for specific layer of considered molecule (core—molecular core; ex. Shell—external layer of the nanomolecule).

Author Contributions: Conceptualization, P.C.; Methodology, P.C.; Software, P.C.; Validation, P.C.; Formal Analysis, P.C.; Investigation, P.C.; Resources, P.C. and B.S.; Data Curation, P.C.; Writing—Original Draft Preparation, P.C.; Writing—Review & Editing, P.C. and B.S.; Visualization, P.C.; Supervision, P.C.; Project Administration, P.C.; Funding Acquisition, P.C.

Funding: This research received no external funding.

Acknowledgments: Thank you to the Professor MV Diudea for sharing the structures. This research was supported by PL-Grid Infrastructure (http://www.plgrid.pl/en).

Conflicts of Interest: The authors declare that there are no conflicts of interest.

References

1. Kunikata, T.; Tatefuji, T.; Aga, H.; Iwaki, K.; Ikeda, M.; Kurimoto, M. Indirubin inhibits inflammatory reactions in delayed-type hypersensitivity. *Eur. J. Pharmacol.* **2000**, *410*, 93–100. [CrossRef]

2.   Kim, S.-A.; Kwon, S.-M.; Kim, J.-A.; Kang, K.W.; Yoon, J.-H.; Ahn, S.-G. 5′-Nitro-indirubinoxime, an indirubin derivative, suppresses metastatic ability of human head and neck cancer cells through the inhibition of Integrin β1/FAK/Akt signaling. *Cancer Lett.* **2011**, *306*, 197–204. [CrossRef] [PubMed]

3.   Williams, S.P.; Nowicki, M.O.; Liu, F.; Press, R.; Godlewski, J.; Abdel-Rasoul, M.; Kaur, B.; Fernandez, S.A.; Chiocca, E.A.; Lawler, S.E. Indirubins decrease glioma invasion by blocking migratory phenotypes in both the tumor and stromal endothelial cell compartments. *Cancer Res.* **2011**, *71*, 5374–5380. [CrossRef] [PubMed]

4.   Shin, E.-K.; Kim, J.-K. Indirubin derivative E804 inhibits angiogenesis. *BMC Cancer* **2012**, *12*, 164. [CrossRef] [PubMed]

5.   Martin, L.; Magnaudeix, A.; Wilson, C.M.; Yardin, C.; Terro, F. The new indirubin derivative inhibitors of glycogen synthase kinase-3, 6-BIDECO and 6-BIMYEO, prevent tau phosphorylation and apoptosis induced by the inhibition of protein phosphatase-2A by okadaic acid in cultured neurons. *J. Neurosci. Res.* **2011**, *89*, 1802–1811. [CrossRef]

6.   Czeleń, P. Molecular dynamics study on inhibition mechanism of CDK-2 and GSK-3β by CHEMBL272026 molecule. *Struct. Chem.* **2016**, *27*, 1807–1818. [CrossRef]

7.   Czeleń, P. Inhibition mechanism of CDK-2 and GSK-3β by a sulfamoylphenyl derivative of indoline—a molecular dynamics study. *J. Mol. Model.* **2017**, *23*, 230. [CrossRef]

8.   Marko, D.; Schätzle, S.; Friedel, A.; Genzlinger, A.; Zankl, H.; Meijer, L.; Eisenbrand, G. Inhibition of cyclin-dependent kinase 1 (CDK1) by indirubin derivatives in human tumour cells. *Br. J. Cancer* **2001**, *84*, 283–289. [CrossRef]

9.   Hoessel, R.; Leclerc, S.; Endicott, J.A.; Nobel, M.E.M.; Lawrie, A.; Tunnah, P.; Leost, M.; Damiens, E.; Marie, D.; Marko, D.; et al. Indirubin, the active constituent of a Chinese antileukaemia medicine, inhibits cyclin-dependent kinases. *Nat. Cell Biol.* **1999**, *1*, 60–67. [CrossRef]

10.   Xiao, Z.; Hao, Y.; Liu, B.; Qian, L. Indirubin and Meisoindigo in the Treatment of Chronic Myelogenous Leukemia in China. *Leuk. Lymphoma* **2002**, *43*, 1763–1768. [CrossRef]

11.   Lee, M.-Y.; Liu, Y.-W.; Chen, M.-H.; Wu, J.-Y.; Ho, H.-Y.; Wang, Q.-F.; Chuang, J.-J. Indirubin-3′-monoxime promotes autophagic and apoptotic death in JM1 human acute lymphoblastic leukemia cells and K562 human chronic myelogenous leukemia cells. *Oncol. Rep.* **2013**, *29*, 2072–2078. [CrossRef]

12.   Czeleń, P.; Szefler, B. Molecular dynamics study of the inhibitory effects of ChEMBL474807 on the enzymes GSK-3β and CDK-2. *J. Mol. Model.* **2015**, *21*, 74. [CrossRef]

13.   De Jong, W.H.; Borm, P.J.A. Drug delivery and nanoparticles:applications and hazards. *Int. J. Nanomedicine* **2008**, *3*, 133–149. [CrossRef]

14.   Gao, Z.; Zhang, L.; Sun, Y. Nanotechnology applied to overcome tumor drug resistance. *J. Control. Release* **2012**, *162*, 45–55. [CrossRef]

15.   Morgen, M.; Bloom, C.; Beyerinck, R.; Bello, A.; Song, W.; Wilkinson, K.; Steenwyk, R.; Shamblin, S. Polymeric Nanoparticles for Increased Oral Bioavailability and Rapid Absorption Using Celecoxib as a Model of a Low-Solubility, High-Permeability Drug. *Pharm. Res.* **2012**, *29*, 427–440. [CrossRef]

16.   Turov, V.V.; Chehun, V.F.; Barvinchenko, V.N.; Krupskaya, T.V.; Prylutskyy, Y.I.; Scharff, P.; Ritter, U. Low-temperature 1H-NMR spectroscopic study of doxorubicin influence on the hydrated properties of nanosilica modified by DNA. *J. Mater. Sci. Mater. Med.* **2011**, *22*, 525–532. [CrossRef]

17.   Schuetze, C.; Ritter, U.; Scharff, P.; Fernekorn, U.; Prylutska, S.; Bychko, A.; Rybalchenko, V.; Prylutskyy, Y. Interaction of N-fluorescein-5-isothiocyanate pyrrolidine-C60 with a bimolecular lipid model membrane. *Mater. Sci. Eng. C* **2011**, *31*, 1148–1150. [CrossRef]

18.   Evstigneev, M.P.; Buchelnikov, A.S.; Voronin, D.P.; Rubin, Y.V.; Belous, L.F.; Prylutskyy, Y.I.; Ritter, U. Complexation of C_{60} Fullerene with Aromatic Drugs. *ChemPhysChem* **2013**, *14*, 568–578. [CrossRef]

19.   Qiao, R.; Roberts, A.P.; Mount, A.S.; Klaine, S.J.; Ke, P.C. Translocation of C_{60} and Its Derivatives Across a Lipid Bilayer. *Nano Lett.* **2007**, *7*, 614–619. [CrossRef]

20.   Prylutska, S.; Bilyy, R.; Overchuk, M.; Bychko, A.; Andreichenko, K.; Stoika, R.; Rybalchenko, V.; Prylutskyy, Y.; Tsierkezos, N.G.; Ritter, U. Water-soluble pristine fullerenes C_{60} increase the specific conductivity and capacity of lipid model membrane and form the channels in cellular plasma membrane. *J. Biomed. Nanotechnol.* **2012**, *8*, 522–527. [CrossRef]

21.   Prylutska, S.V.; Grynyuk, I.I.; Grebinyk, S.M.; Matyshevska, O.P.; Prylutskyy, Y.I.; Ritter, U.; Siegmund, C.; Scharff, P. Comparative study of biological action of fullerenes C_{60} and carbon nanotubes in thymus cells. *Materwiss. Werksttech.* **2009**, *40*, 238–241. [CrossRef]

22. Johnston, H.J.; Hutchison, G.R.; Christensen, F.M.; Aschberger, K.; Stone, V. The Biological Mechanisms and Physicochemical Characteristics Responsible for Driving Fullerene Toxicity. *Toxicol. Sci.* **2010**, *114*, 162–182. [CrossRef]

23. Andrievsky, G.; Klochkov, V.; Derevyanchenko, L. Is the C_{60} Fullerene Molecule Toxic?! *Fuller. Nanotub. Carbon Nanostructures* **2005**, *13*, 363–376. [CrossRef]

24. Satoh, M.; Takayanagi, I. Pharmacological studies on fullerene (C_{60}), a novel carbon allotrope, and its derivatives. *J. Pharmacol. Sci.* **2006**, *100*, 513. [CrossRef]

25. Yin, J.-J.; Lao, F.; Fu, P.P.; Wamer, W.G.; Zhao, Y.; Wang, P.C.; Qiu, Y.; Sun, B.; Xing, G.; Dong, J.; et al. The scavenging of reactive oxygen species and the potential for cell protection by functionalized fullerene materials. *Biomaterials* **2009**, *30*, 611–621. [CrossRef]

26. Partha, R.; Conyers, J.L. Biomedical applications of functionalized fullerene-based nanomaterials. *Int. J. Nanomedicine* **2009**, *4*, 261–275.

27. Echalier, A.; Hole, A.J.; Lolli, G.; Endicott, J.A.; Noble, M.E.M. An inhibitor's-eye view of the ATP-binding site of CDKs in different regulatory states. *ACS Chem. Biol.* **2014**, *9*, 1251–1256. [CrossRef]

28. Simon, J.; Flahaut, E.; Golzio, M. Overview of Carbon Nanotubes for Biomedical Applications. *Materials* **2019**, *12*, 624. [CrossRef]

29. Lucente-Schultz, R.M.; Moore, V.C.; Leonard, A.D.; Price, B.K.; Kosynkin, D.V.; Lu, M.; Partha, R.; Conyers, J.L.; Tour, J.M. Antioxidant Single-Walled Carbon Nanotubes. *J. Am. Chem. Soc.* **2009**, *131*, 3934–3941. [CrossRef]

30. Czeleń, P.; Szefler, B. The Immobilization of Oxindole Derivatives with Use of Cube Rhombellane Homeomorphs. *Symmetry* **2019**, *11*, 900. [CrossRef]

31. Szefler, B.; Czeleń, P. Docking of Cisplatin on Fullerene Derivatives and Some Cube Rhombellane Functionalized Homeomorphs. *Symmetry* **2019**, *11*, 874. [CrossRef]

32. Szefler, B.; Czeleń, P.; Diudea, M.V. Docking of indolizine derivatives on cube rhombellane functionalized homeomorphs. *Stud. Univ. Babes-Bolyai Chem.* **2018**, *63*, 7–18. [CrossRef]

33. Diudea, M.V.; Lungu, C.N.; Nagy, C.L.; Diudea, M.V.; Lungu, C.N.; Nagy, C.L. Cube-Rhombellane Related Structures: A Drug Perspective. *Molecules* **2018**, *23*, 2533. [CrossRef]

34. CHEMBL database release 24.1. 2018. Available online: https://www.ebi.ac.uk/chembl/ (accessed on 2 February 2014).

35. Kim, K.-H.; Ko, D.K.; Kim, Y.-T.; Kim, N.H.; Paul, J.; Zhang, S.-Q.; Murray, C.B.; Acharya, R.; Kim, Y.H.; DeGrado, W.F.; et al. Protein-directed self-assembly of a fullerene crystal. *Nat. Commun.* **2016**, *7*, 11429. [CrossRef]

36. PubChem. Available online: https://pubchem.ncbi.nlm.nih.gov/ (accessed on 5 May 2019).

37. Humphrey, W.; Dalke, A.; Schulten, K. VMD: Visual molecular dynamics. *J. Mol. Graph.* **1996**, *14*, 33–38. [CrossRef]

38. Trott, O.; Olson, A.J. AutoDock Vina: Improving the speed and accuracy of docking with a new scoring function, efficient optimization, and multithreading. *J. Comput. Chem.* **2010**, *31*, 455–461. [CrossRef]

39. Bartashevich, E.V.; Potemkin, V.A.; Grishina, M.A.; Belik, A.V. A Method for Multiconformational Modeling of the Three-Dimensional Shape of a Molecule. *J. Struct. Chem.* **2002**, *43*, 1033–1039. [CrossRef]

Article

Investigation of the Inhibition Potential of New Oxindole Derivatives and Assessment of Their Usefulness for Targeted Therapy

Przemysław Czeleń

Department of Physical Chemistry, Faculty of Pharmacy, Collegium Medicum, Nicolaus Copernicus University, Kurpinskiego 5, 85-096 Bydgoszcz, Poland; przemekcz@cm.umk.pl

Received: 29 June 2019; Accepted: 27 July 2019; Published: 1 August 2019

Abstract: Oxindole derivatives are a large group of compounds that can play the role of Adenosine triphosphate (ATP) competitive inhibitors. The possibility of modification of such compounds by addition of active groups to both cyclic systems of oxindole allows the obtaining of derivatives showing significant affinity toward cyclin-dependent kinase (CDK) proteins. Overexpression of that enzyme is observed in the case of most cancers. The discovery of new efficient inhibitors, which could be used in the development of targeted therapies, is one of the current goals setting trends in recent research. In this research, an oxindole molecular core was used, which was modified by the addition of different substituents to both side chains. The realized procedure allowed the creation of a set of oxindole derivatives characterized by binding affinity values and molecular descriptors evaluated during docking procedures and QSAR calculations. The most promising structures characterized by best sets of parameters were used during the molecular dynamics stage. The analysis of structural and energetic properties of systems obtained during this stage of computation gives an indication of inhibitors creating the most stable complexes, characterized by the highest affinity. During this stage, two structures were selected, where affinity towards potential nanocarriers was evaluated. Realized calculations confirmed a significant role of stacking interactions in the stabilization of ligand complexes with fullerene molecules. Obtained data indicates that complexes of oxindole derivatives and considered nanocarriers exhibit significant potential in the creation of immobilized drugs, and can be used in the development of targeted therapies.

Keywords: CDK-2; oxidoles; docking; molecular dynamics; MMPBSA; inhibition; C_{60} fullerene derivatives

1. Introduction

The process of cell proliferation is crucial for all living organisms. The regulation of the cell cycle is strictly related to the action of a set of SER/THR kinases classified as cyclin-dependent kinases (CDKs). All these proteins, after complexation with activate factors called cyclins, play the role of regulation of activity of other proteins involved in transcription and replication processes [1–3]. Among all CDK proteins the most crucial role in progression of the cell cycle is assigned to CDK2. Overexpression of that enzyme is often observed in the case of most cancers [2,4–7]. The search for the ATP competitive inhibitors of CDK2 is important in the context of creation of anticancer therapies. The example of a natural product exhibiting such properties is indirubin [8,9] the anticancer properties of this compound were widely known and used in traditional Chinese medicine. The chemical structure of the indirubin molecule consists of two conjugated oxindoles. The derivatives and analogs of indirubine represent a large group of new CDK2 inhibitors [10–14]. The oxindole core was repeatedly the basis of the creation of ATP competitive inhibitors. Luk et al showed that the addition of active groups to both cyclic systems of oxindole allows the obtaining of derivatives showing significant affinity toward the CDK2 active site, which can also exhibit selective potential in the context of interactions with proteins from the CDK family [15,16]. The CDK2 inhibitors analyzed in earlier works [17,18] confirm that the oxindole core with a carbonyl group substituted in C5 position exhibits high affinity toward the CDK2 active site. The presence of hydrogen-bond donors and acceptors in this structure provides strong anchorage in the space of the active site through characteristic interaction with Glu 81, Leu 83, and Lys 33 [17,18] (see Figure 1). In this work, it was proposed to use such a structure as the basic core of new oxindole derivatives created by the addition of a set of different active groups exhibiting diverse binding properties (see Figure 2). The screening of obtained structures, including binding affinity and molecular properties, allows for the prediction of new ATP competitive inhibitors well suited to the structural and binding properties of the CDK2 active site. An important aspect of drug development is also the finding of effective delivery methods. One commonly used technique in the creation of targeted therapies is immobilization of drugs with the use of different types of carriers [19–21], which ensures their direct delivery and release to the appropriate biological target. The use of such a method has other potential related effects, e.g., decreasing drug toxicity [22,23], and increasing its bioavailability and elongation of pharmacological action [24,25]. One of the most commonly used nanomolecules employed for such an aim is C_{60} fullerene and its derivatives [23,26–29]. Such a molecule exhibits high affinity towards biologically active molecules containing aromatic systems. This popular fullerene exhibits biological activity towards cells [27], its characteristic chemical structure enables good permeation trough cell membrane [30,31], and small concentrations of this compound are nontoxic for living organisms. The investigation of oxindole derivative binding potential relative to C_{60} and its derivatives allows evaluation of the possibilities of their use in the creation of targeted therapies.

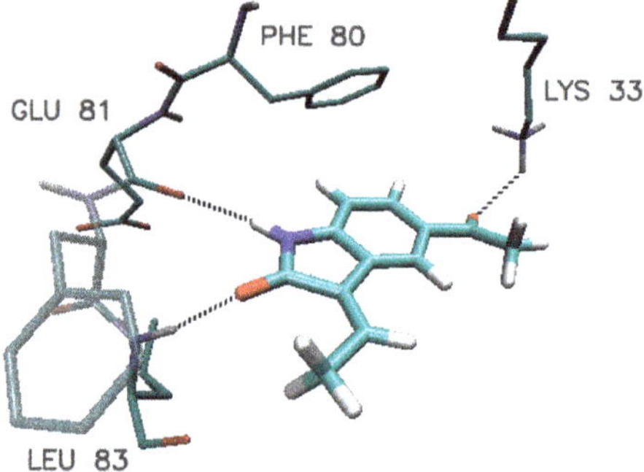

Figure 1. The hydrogen bonds and stacking interactions characteristic for oxindole core with amino acids from CDK2 active site.

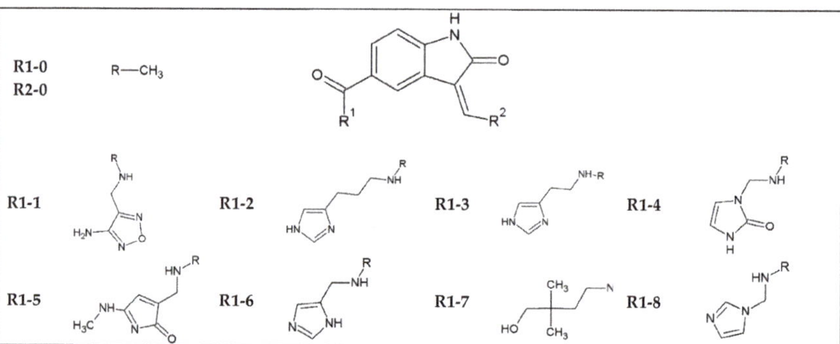

Figure 2. *Cont.*

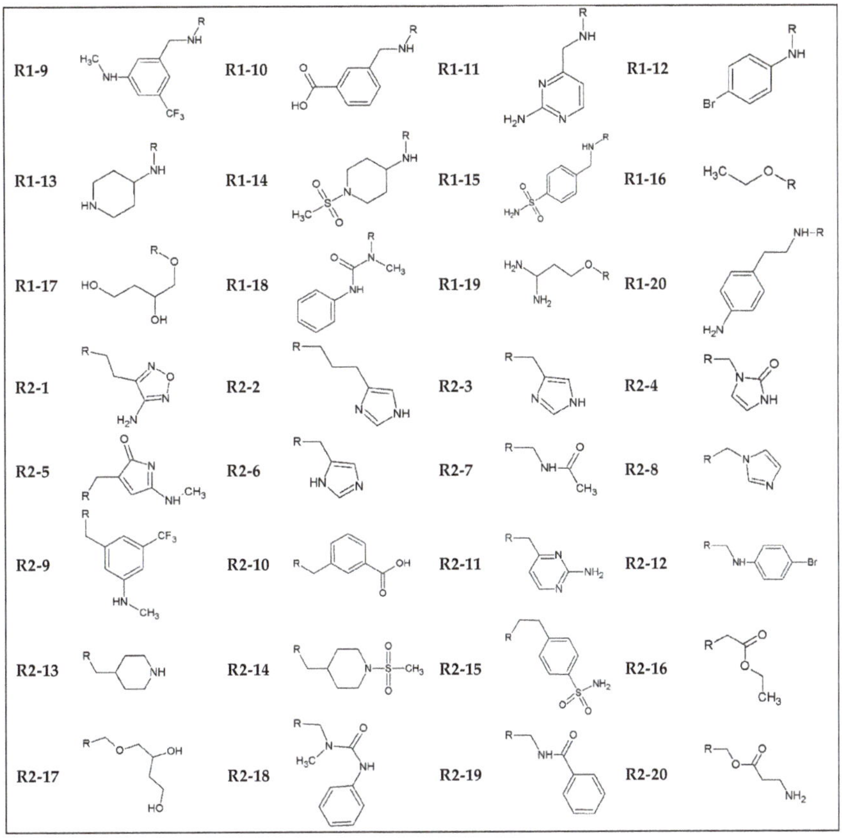

Figure 2. The graphic representation of oxindole core and active groups used during the creation of ligands.

2. Methods

Ligand structures were created by the addition of 20 different active groups to both side chains of the oxindole core. In the first step, only one side chain was replaced by an active group, while the second was represented by a methyl group. In this way, such active groups were determined, which increase binding affinity of ligand towards the CDK2 active site with the preservation of interactions characteristic for indoline derivatives identified in previous works [17,18]. A total of 84 different indoline derivatives were created with the use of seven active groups in R1 position and 14 in R2. The designation of the structures created in this procedure contains the prefix (Indol_) and the numbers of the bases used in positions R1 and R2. All considered structures were optimized during computations with the use of TURBOMOL at COSMO-BP-tzvpd-fine level of theory [32]. The toxicity of obtained molecules was simulated using a combination of the 3D/4D QSAR (Quantitative structure–activity relationship) BiS/MC and CoCon algorithms [33,34] with the use of the Chemosophia application [35]. The octanol–water partition coefficient (LogP) was determined with use of the COSMOTHERM application [36]. All docking procedures were realized with the use of AutoDockVina [37]. The C_{60} structure and cyclin-dependent kinase 2 (CDK2-PDB ID 1E9H) were downloaded from Brookhaven Protein Database PDB, while C_{60} functionalized derivatives were obtained from PubChem database [38]. All calculations were realized with the use of chemical structures of ligands, protein, and nanocarriers containing only polar hydrogen atoms. All initial steps were realized with the use of the AutoDock

Tolls package [39]. To confirm the location of the active site and to exclude the presence of competitive interactions on the surface of the protein, blind docking was performed [40,41]. The size of the grid box was 72 × 84 × 66 Å and covered the entire surface of the protein. The realized preliminary calculations confirmed the localization of the active site and the dimensions of the grid box were fitted to the size of active site of CDK2 enzyme (16 × 14 × 22 Å). In the case of all nanoparticles, the grid box dimensions were established equal to 26 × 26 × 26 Å. During the molecular dynamics procedure, there were structures of complexes created by CDK-2 protein with chosen ligand molecules. Ligand structures were characterized using generalized amber force-field parameters. In the case of the CDK-2 protein, the ff14SB Force Field [42] was used and the atomic charges were calculated according to the Merz–Kollmann scheme via the RESP (Restrained electrostatic potential atomic partial charges) procedure at HF/6-31G* level [43]. Each system was neutralized with the use of three chloride anions and immersed in a periodic box (79 x 86 x 71 Å) consisting of 14,809 TIP3P water molecules. Considered systems were heated to 300K by 100 ps of initial MD simulation, while the temperature was controlled by Langevin thermostat [44]. The minimization and heating of systems was realized with constant-volume periodic boundaries (NTB = 1), while the proper production of molecular dynamics was realized with constant pressure (NTB = 2). The periodic boundary conditions and SHAKE algorithm were applied for 80 ns of molecular dynamic simulation; the first 20 ns of simulation time was used as equilibration interval, and the next 60 ns of the trajectory were used in the analysis of the interaction between the considered subunits. Structural analysis was performed with use of the VMD package [45]. The energetic characteristic of interaction between the ligand and the active site was obtained with the use of Molecular Mechanic/Poisson–Boltzmann Surface Area (MMPBSA) method [46]. In all molecular dynamics simulations, the AMBER 14 package was used [47]. During analysis of interactions in the active site, the hydrogen bonds were defined by the following criteria: the distance between donor (D) and acceptor (A) < 3.5 Å, angle D–H–A > 90°, and distance H–A < 3Å.

3. Results and Discussion

The structures of ligand molecules obtained by the addition of a single functional group to the oxindole core of the molecule were docked to the CDK2 active site. Each obtained complex preserved characteristic interactions between the oxindole core and the CDK2 active site, namely with LEU 83, GLU 81, and LYS 33. The binding affinities of considered structures toward protein are presented in Table 1. The modifications applied to the R1 position in 16 cases increased the binding affinity of the ligand toward the CDK2 active site. The highest observed increase of this value did not exceed 16% and only in seven cases exceeded 10%. All active groups exhibiting the best impact on binding properties of ligands contain in their structure aromatic or heterocyclic rings and chemical groups which can fulfil the role of hydrogen-bond donors or acceptors. Among all considered chemical groups, only in four cases was there observed a decrease of binding affinity of ligand toward protein, each of them being an aliphatic system with hydrogen-bond donors. The use of a similar set of active groups in the R2 position meant that for 18 ligand structures there was observed an increase of affinity toward the CDK2 active site. In 14 cases it exceeded 10% and the highest reported values indicated almost a 25% increase of binding properties. The highest impact on the increase of binding affinity has substitution of active groups number 10, 9, 15, 1, 11, 13, and 12 in R2 position. The comparison of energy values characterizing all modified structures indicates that substitutions in the R2 position have a greater impact on binding affinity than modification added to R1. The next step of docking proceedings used ligand structures created with the use of active groups, which ensured at least 10% increase of binding affinity (7 substitutions R1, 14 substitutions R2). The values describing binding affinity and basic molecular descriptors for the chosen ligand molecules are presented in Table 2, and the characteristics of all structures analyzed in this work are placed in the Appendix A Table A1. During ligand selection, the used criteria encompassed not only the values of binding affinity and related inhibition constant (IC), but also molecular descriptors such as LogP and toxicity of considered molecules. The toxicity of the analyzed compounds estimated during QSAR predictions is described by values in the range from 0 to 1. The compounds with very low toxicity adopt values from 0.0–0.2, the moderate toxic ones correspond to values 0.2–0.8, and highly toxic compounds to values 0.8–1.0, with the standard error for estimated values being 0.1. Among all considered oxindole derivatives, only 23 structures were characterized by values indicating nontoxic character. In the next stage, compounds characterized as medium toxic, with toxicity index below 0.4, were also used, which allowed an increase in the studied group to 40 compounds. The IC for ligand molecules was estimated based on the equation

$$K_I = exp^{\left(\frac{\Delta G_b}{RT}\right)}$$

where ΔG_b represents binding affinity, R gas constant, and T temperature. The analysis including all presented factors allows for selection of potential inhibitors characterized by small values of inhibition constant, slight toxicity, and potentially good permeability through cell membranes (LogP). The best obtained structures are characterized by ~30% increase of binding affinity toward the CDK2 binding site, which is also related to significant changes in IC, which decreased in best cases to values lower than 10 nM (IC for Indol_0_0 = 824.2 nM).

Table 1. The values of biding affinity of oxindol mono derivatives towards CDK2. Increase of binding affinity estimated relative to the value for Indol_0_0 equals −8.3 kcal/mol.

Name	Binding Affinity (kcal/mol)	Increase of Binding Affinity (%)	Name	Binding Affinity (kcal/mol)	Increase of Binding Affinity (%)
Indol_9_0	−9.6	15.7	Indol_0_10	−10.3	24.1
Indol_15_0	−9.5	14.5	Indol_0_9	−10.2	22.9
Indol_10_0	−9.4	13.3	Indol_0_15	−10.1	21.7
Indol_20_0	−9.3	12.0	Indol_0_11	−9.8	18.1
Indol_1_0	−9.3	12.0	Indol_0_1	−9.66	16.4
Indol_4_0	−9.24	11.3	Indol_0_13	−9.6	15.7
Indol_18_0	−9.22	11.1	Indol_0_12	−9.58	15.4
Indol_11_0	−9	8.4	Indol_0_4	−9.3	12.0
Indol_8_0	−8.9	7.2	Indol_0_6	−9.3	12.0
Indol_6_0	−8.9	7.2	Indol_0_19	−9.3	12.0
Indol_3_0	−8.9	7.2	Indol_0_3	−9.2	10.8
Indol_13_0	−8.9	7.2	Indol_0_5	−9.2	10.8
Indol_14_0	−8.86	6.7	Indol_0_8	−9	8.4
Indol_2_0	−8.8	6.0	Indol_0_14	−9	8.4
Indol_12_0	−8.76	5.5	Indol_0_2	−8.96	8.0
Indol_5_0	−8.74	5.3	Indol_0_18	−8.82	6.3
Indol_19_0	−8.2	−1.2	Indol_0_7	−8.8	6.0
Indol_17_0	−8.2	−1.2	Indol_0_16	−8.5	2.4
Indol_7_0	−8.02	−3.4	Indol_0_20	−8.3	0.0
Indol_16_0	−7.92	−4.6	Indol_0_17	−8.24	−0.7

The interactions involved in the stabilization of complexes of five chosen ligands exhibiting best properties are presented in Figure 3; the summary presentation of all observed interactions is attached in Table 3. For all obtained ligand molecules, characteristic interactions of oxindole derivatives created by atoms from molecule core are observed. Considering geometric classification of hydrogen-bond strength, interactions with Leu 83, Glu 81, and Lys 33 can be classified as medium-strength impacts in the case of all chosen structures, while the bonds created with Asp 146 are weak. For all considered ligands, potential stacking interactions with aromatic system of Phe 80 were also observed, and in all cases similar distances between aromatic rings and their analogical mutual orientation were noticed. The addition of active groups in the R1 and R2 positions provided additional impacts, which are hydrogen bonds with Gln 85 (all), Glu 12 (Indol_4_9; Indol_4_10), Asp 86 (Indol_20_10, Indol_20_15, Indol_4_9), Lys 89 (Indol_4_10, Indol_20_15, Indol_9_15). In all cases they can be classified as medium- or weak-strength interactions.

Table 2. The values of biding affinity and molecular parameters for the structures exhibiting the best properties.

	Name	LogP	Toxicity	Binding Affinity (kcal/mol)	Increase of Binding Affinity (%)	Inhibition Constant (nM)
1.	indol_20_10	3.61	0.39	−11.20	34.94	6.17
2.	indol_4_10	2.04	0.07	−11.10	33.73	7.30
3.	indol_4_9	3.58	0.32	−11.00	32.53	8.65
4.	indol_9_15	3.64	0.19	−10.88	31.08	10.59
5.	indol_20_15	2.18	0.19	−10.80	30.12	12.12
6.	indol_18_12	3.85	0.28	−10.54	26.99	18.80
7.	indol_18_15	1.65	0.00	−10.44	25.78	22.25
8.	indol_10_4	1.58	0.08	−10.44	25.78	22.25
9.	indol_18_10	2.98	0.03	−10.40	25.30	23.81
10.	indol_20_4	1.46	0.11	−10.40	25.30	23.81
11.	indol_4_13	1.41	0.17	−10.40	25.30	23.81
12.	indol_15_12	1.95	0.30	−10.40	25.30	23.81
13.	indol_4_12	2.80	0.35	−10.40	25.30	23.81
14.	indol_20_1	1.26	0.39	−10.40	25.30	23.81
15.	indol_10_13	3.38	0.28	−10.36	24.82	25.47

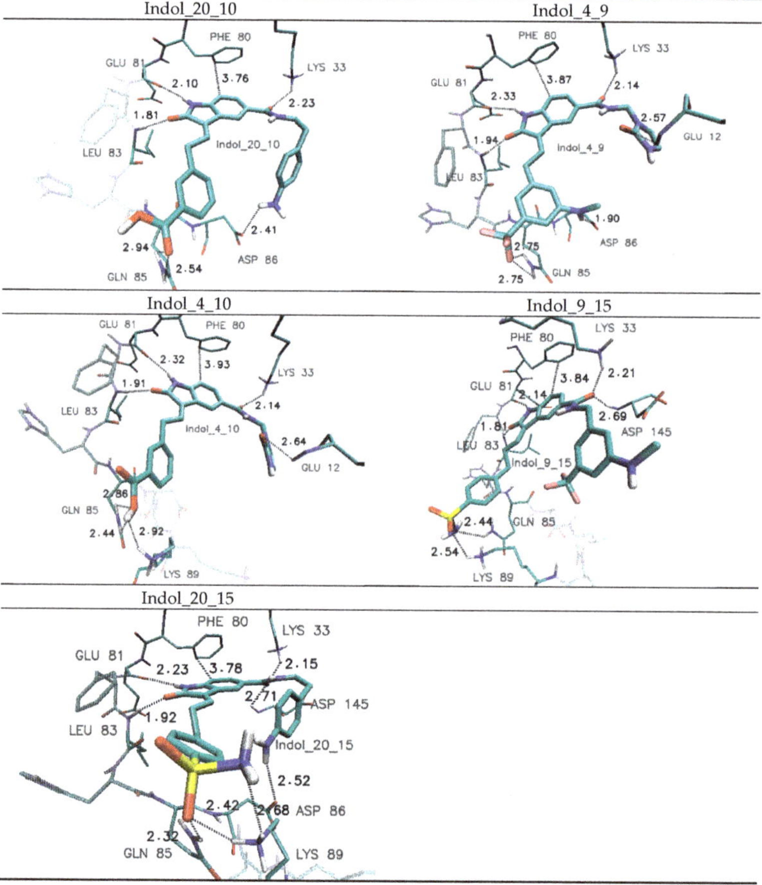

Figure 3. The graphic representations of interactions of chosen oxindole derivatives with CDK2 active site identified during docking stage.

Table 3. Interactions stabilizing oxindole derivatives complexes with CDK2 active site, estimated during docking simulation stage.

Amino Acid	Hydrogen-Bond Length/Distance Between Aromatic Systems * (Å)				
	Indol_4_9	Indol_4_10	Indol_9_15	Indol_20_10	Indol_20_15
Glu 12	2.57	2.64	—	—	—
Lys 33	2.14	2.14	2.21	2.23	2.15
Leu 83	1.94	1.91	1.81	1.81	1.92
Glu 81	2.33	2.32	2.14	2.10	2.23
Gln 85	2.75	2.44 2.86	2.44	2.54 2.94	2.32
Asp 86	1.90	—	—	2.41	2.52
Lys 89	—	2.92	2.54	—	2.42 2.68
Asp146	2.86	2.86	2.69	2.61	2.71
Phe 80 *	3.87	3.87	3.84	3.76	3.84

4. Molecular Dynamics of Chosen Complexes

The molecular dynamics stage of simulations covered five complexes of CDK2 protein with selected ligands characterized by best binding and molecular properties. The dynamic and structural stability of the considered systems was evaluated with the use of root mean square deviation (RMSD) values, which where estimated for protein and inhibitor molecules. The appropriate values describing each element of the analyzed complexes are presented in Figure 4 and Table 4. Each presented system reached structural equilibrium in the first 20 ns stages of the molecular dynamics simulation. The distributions and averaged values presented in the table show that CDK2 protein exhibits similar structural and dynamic properties in each considered system. The behavior of the analyzed inhibitor molecules in the protein active site is more varied. The highest structural stiffness is observed in the case of Indol_4_9 and Indol_20_10 molecules; small average values and uniform distributions indicate stabile structural conformations of both inhibitors. The differences in standard deviations describing both populations are related to incidental structural fluctuations observed in the case of Indol_20_10. The rest of the considered ligand molecules are characterized by more diverse distributions. In the case of Indol_4_10, values presented on the chart indicate the potential presence of a second conformation, which appears after 30 ns of simulation time. The structures containing sulfonamide group in R2 position (Indol_9_15, Indol_20_15) exhibit higher structural flexibility than the rest of considered inhibitors, which confirms the observed fluctuations and higher standard deviations of RMSD values. The observed general trends of ligands molecule dynamic properties are also reflected in the durability of interactions involved in the stabilization of the analyzed complexes. The data presented in Table 5 describes the quantity and quality of the interactions reported for complexes during molecular dynamics simulations. Not all interactions identified during the molecular docking stage proved to be stable; however, in some cases new impacts appeared to be created with other amino acids. All realized molecular dynamics simulations revealed that interactions with Leu 83 and Glu 81 play a crucial role in the stabilization of inhibitor-CDK2 complexes. Such hydrogen bonds were observed for all conformations collected for each ligand during computations; the bond distances indicate that most can be classified as medium-strength interactions (~80% for Leu 83, ~75% for Glu 81). The third interaction created by atoms from the molecular core of the considered ligands, namely with Lys 33, is characterized by a much larger diversity, which manifests in the quantity and strength of created bonds alike. The most stable interactions are observed in the case of the Indol_4_9 molecule; however, the largest share of interactions with the highest impact energy (~40%) was recorded for Indol_9_15 and Indol_20_15 molecules. The weak interactions reported for Asp 146 disappeared during the molecular dynamics simulation. Much more important discrepancies in relation to initial complex conformations are observed for interactions of groups placed in the side chains of considered ligands. The first considered ligand, namely Indol_4_9, creates a stabile interaction with oxygen from aspartic acid ASP 86; such a hydrogen bond is observed in 98% of conformations obtained during molecular dynamics simulation. The other impacts identified during the docking procedure have changed, namely weak interactions with GLU 12 are observed only for 45% of collected conformations, while hydrogen bonds with GLN 85 have disappeared. In the case of Indol_4_10, two hydrogen bonds created by groups from side chains were observed. The first of them, created with GLU12, is observed for 46.4% of conformations, while the second impact, namely the hydrogen bond with GLU8 (69.1%), does not occur in the complex obtained during the docking phase. The appearing of this new interaction is a consequence of the change of R2 side chain conformation, which is also related to the disappearance of hydrogen bonds with LYS 98 and GLN 85. All the hydrogen bonds found for active groups from the side chains of Indol_9_15 during docking stage disappeared. The analysis of possible interactions for conformers obtained during molecular dynamics stage indicates only one hydrogen bond with GLY13; however, the quantity and distances indicate a weak stabilizing impact of this bond. The next molecule, namely Indol_20_10, interacts with the active site by hydrogen bonds created by both side chains. The hydrogen bond identified between the amino group of side chain R1 and ASP 86 is observed for 50% of conformations obtained during molecular dynamics calculations. The structural mobility of this part of ligand molecule caused the occurrence of a second competitive interaction created with the oxygen atom of ILE10 (61.3%). The active group from the second side chain also creates a bond, namely with GLU 8 (85.7%); the quantity and distance of this hydrogen bond clearly indicate a significant stabilizing impact for the complex structure. In the case of the last considered ligand, namely Indol_20_15, during the molecular dynamics stage the presence of two hydrogen bonds observed in initial complex and one new impact was identified and confirmed. Both oxygens from the sulfonamide group are involved in interactions with LYS89 (73.6%); the rotation of this group relative to the aromatic ring allowed for the appearance of a new bond observed between amide hydrogens and oxygen from HIE 84 (56.7%). The activity of the second side chain is much less significant; the hydrogen bond between amine group and ASP 86 is observed only for 48% of the conformations. The values of binding enthalpies (ΔH) obtained for complexes considered during the molecular dynamic stage strictly correspond to the dynamic properties of ligands described by RMSD values, as well as the quantity and quality of interactions involved in the stabilization of complexes. Values presented in Table 6 clearly show that the most stable complexes during molecular dynamics stage were created by two inhibitors, namely Indol_4_9 and Indol_20_10, with very similar values of enthalpy (~40 kcal/mol) and standard deviations indicating a similar inhibiting potential of these two compounds. The next group of ligands are Indol_4_10 and Indol_20_15, which are characterized by enthalpy values about 6 kcal/mol smaller than those obtained for the best inhibitors. The worst inhibiting potential among all structures analyzed during the molecular dynamics stage is shown by the Indol_9_15 molecule.

Table 4. The averaged values of RMSD (Å) for ligands and enzyme for all steps used during structural analysis. The values in italics represent standard deviations.

	Indol_4_9		Indol_4_10		Indol_9_15		Indol_20_10		Indol_20_15	
	CDK2	LIG	CDK2	LIG	CDK2	LIG	CDK2	LIG	CDK2	LIG
RMSD	2.36	1.30	2.91	1.54	2.46	1.68	2.72	0.93	2.49	1.46
SD	0.10	0.08	0.17	0.26	0.15	0.27	0.13	0.19	0.14	0.29

Table 5. Distribution of the most frequently created hydrogen bonds between ligands molecule and selected amino acids from CDK-2 active site. The hydrogen bonds in the table represent median values of intervals with a width of 0.25 Å.

Hydrogen Bond	Population %							
	Σ	1.5 Å	1.75 Å	2.0 Å	2.25 Å	2.5 Å	2.75 Å	3.0 Å
Indol_4_9								
Ligand (O2) ... (HN) LEU 83	100.0	0.0	24.1	58.9	14.8	2.1	0.1	0.0
Ligand (H3) ... (O) GLU 81	100.0	0.0	14.1	58.3	24.1	3.3	0.3	0.0
Ligand (O3) ... (H) LYS 33	98.7	0.0	24.4	44.4	16.0	8.5	2.9	2.4
Ligand (H1) ... (O) ASP 86	98.3	0.0	10.2	37.1	29.4	13.7	5.5	2.4
Ligand (H4) ... (O) GLU 12	45.4	0.0	0.1	1.4	3.6	7.9	12.7	19.7
Indol_4_10								
Ligand (O2) ... (HN) LEU 83	100.0	0.1	31.1	55.1	11.4	2.1	0.2	0.0
Ligand (H3) ... (O) GLU 81	100.0	0.0	19.5	57.2	19.1	3.3	0.9	0.0
Ligand (O3) ... (H) LYS 33	82.9	0.1	26.9	35.8	12.1	5.6	1.2	1.3
Ligand (H2) ... (O) GLU 8	69.1	15.4	43.6	8.4	1.0	0.4	0.1	0.2
Ligand (H4) ... (O) GLU 12	46.4	0.0	10.0	21.1	8.5	3.3	2.3	1.3
Indol_9_15								
Ligand (O3) ... (HN) LEU 83	100.0	0.1	24.5	55.2	17.0	2.3	0.7	0.3
Ligand (H5) ... (O) GLU 81	100.0	0.1	27.1	57.8	12.9	1.9	0.3	0.0
Ligand (O4) ... (H) LYS 33	94.3	0.2	44.3	40.5	7.3	1.5	0.3	0.3
Ligand (F) ... (H) GLY 13	31.4	0.0	0.0	0.3	1.0	3.8	9.6	16.8
Indol_20_10								
Ligand (O1) ... (HN) LEU 83	100	0.1	25.7	57.3	14.8	2.3	0.0	0.0
Ligand (H5) ... (O) GLU 81	100	0.0	27.9	56.6	13.5	1.9	0.2	0.0
Ligand (O4) ... (H) LYS 33	92.5	0.1	38.9	35.3	13.8	2.0	1.0	1.4
Ligand (NH$_\chi$) ... (O) ASP 86	50.7	0.2	18.8	22.1	5.2	1.6	1.1	1.7
Ligand (H4) ... (O) GLU 8	85.7	21.3	54.2	8.8	1.0	0.3	0.1	0.1
Ligand (NH$_\chi$) ... (O) ILE 10	61.3	0.0	10.9	25.0	11.2	4.5	4.9	4.8
Indol_20_15								
Ligand (O3) ... (HN) LEU 83	100.0	0.1	23.2	59.4	15.0	2.3	0.1	0.0
Ligand (H6) ... (O) GLU 81	100.0	0.0	24.3	61.4	13.2	0.9	0.1	0.0
Ligand (O4) ... (H) LYS 33	89.6	0.3	38.4	39.4	8.2	2.5	0.6	0.2
Ligand (H3) ... (O) ASP 86	48.0	0.0	11.4	20.0	9.0	3.3	2.6	1.7
Ligand (H1) ... (O) HIE 84	56.7	0.1	20.9	25.1	6.8	1.6	0.9	1.2
Ligand (O1/2) ... (H) LYS 89	73.6	0.1	6.7	23.5	17.6	11.8	8.5	5.4

Table 6. The values of binding enthalpy (kcal/mol) estimated for all complexes considered during molecular dynamics stage ($E_{VDWAALS}$ = van der Waals contribution from MM.; E_{EL} = electrostatic energy; E_{PB} = the electrostatic contribution to the solvation free energy calculated by PB; E_{CAVITY} = nonpolar contribution to the solvation free energy; ΔH = final estimated binding enthalpy).

	Indol_4_9		Indol_4_10		Indol_9_15		Indol_20_10		Indol_20_15	
	ΔE	SD	ΔE	SD	ΔE	SD	ΔE	SD	ΔE	SD
$E_{VDWAALS}$	−49.11	2.69	−46.55	4.11	−52.15	3.91	−49.56	5.35	−48.56	7.74
E_{EL}	−51.37	5.47	−48.50	11.03	−51.08	9.87	−48.47	8.36	−47.83	12.99
E_{PB}	64.27	6.44	65.24	8.06	77.96	8.56	63.82	5.35	70.35	11.03
E_{CAVITY}	−4.47	1.04	−4.65	1.70	−4.22	2.20	−5.11	0.19	−6.98	3.60
ΔH	−40.67	8.81	−34.47	9.02	−29.48	10.32	−39.31	8.31	−33.02	7.22

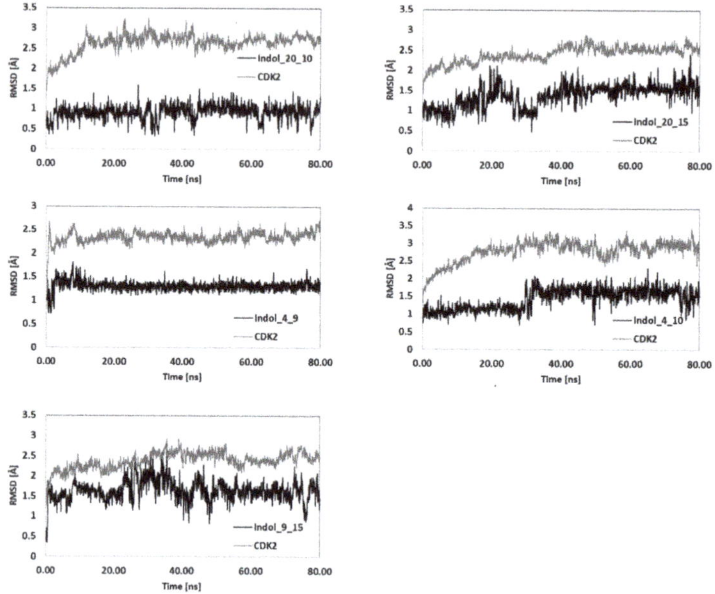

Figure 4. Distributions of RMSD (Å) values. Black distributions refer to ligand molecules while gray distributions refer to CDK-2 protein.

5. The Immobilization of Chosen Ligands with the Use of C_{60} Fullerene Derivatives

In the last stage of research, two ligand molecules were used which exhibit the highest affinity towards active site during molecular dynamics stage, namely Indol_4_9 and Indol_20_10. The affinity of the mentioned molecules towards C_{60} fullerene and its derivatives was evaluated during docking calculations. The data presented in Tables 7 and 8 show that both considered ligands exhibit similar affinity toward native C_{60} fullerene; however, slightly higher values are denoted for the Indol_20_10 molecule. Both considered ligand molecules contain in each of their side chains one cyclic system, aromatic or heterocyclic, and such a chemical structure allows an occurrence of stacking interactions. The graphic representations of these complexes included in Figure 5 confirm that all cyclic systems of both inhibitors are involved in such a type of interaction. The planar orientation of all cyclic systems relative to fullerene surface and relatively small distances placed in the range from 3.36 to 3.77 Å indicate an occurrence of strong stacking interactions stabilizing the considered complexes. The rest of the values presented in Tables 7 and 8 concerns functionalized derivatives of C_{60} fullerene. Each of these structures was created by a single or multiple addition of functional groups to the fullerene surface. Among all used substituents there can be found aliphatic chains with halogen, methoxy, ester, and amide groups, and aromatic systems such as unsubstituted benzene rings and more complex aromatic systems. The description of all used nanosystems is presented in Table 2. All the structural modifications observed in the considered nanocarriers contributed to an increase of binding affinities towards both ligand molecules. In the case of Indol_4_9, the observed differences are placed in a range from −0.7 to −1.8 kcal/mol while for Indol_20_10 from −0.8 to −2 kcal/mol. Much more evident discrepancies are observed in the case of binding constant for the complex creation. The comparison of the best obtained complexes relative to the reference system (C_{60} fullerene) indicates an increase in the yield of the complex formation reaching as much as 2824%. The Indol_4_9 exhibits the highest binding affinity toward FF_12 fullerene, and the increase of yield of this complex formation relative to the reference system is equal to 1986.5%. The complex obtained during the docking stage is maintained by several stacking interactions occurring between all cyclic systems of ligand and surface of fullerene;

observed distances are placed in the range from 3.26 to 3.58 Å. Also, the orientation of both cyclic systems from side chains towards aromatic substituents placed on the fullerene surface indicates the presence of additional stabilizing impacts. The next interesting complexes of this ligand are created with FF_5 and FF_11 nanomolecules; both are characterized by similar affinity values. In both complexes, all cyclic systems are involved in the creation of stacking interactions. In the case of complexes with the FF_5 molecule, a new stabilizing impact appears, namely the hydrogen bond between the hydrogen from the amine group and oxygen from the phosphoryl group with the distance of 1.96 Å, indicating the medium strength of this impact. The complexes created by Indol_20_10 are characterized by higher affinity values in relation to the previously discussed inhibitor. The most stable complexes are created with FF_12 and FF_11 nanomolecules and the value of binding affinity equals −9.2 kcal/mol, causing the yield of these complexes creation to be higher by 2824% relative to the reference system. The graphic representations presented in Figure 5 confirm that in both complexes all aromatic systems of the considered ligand create stacking interactions with fullerene surface, and in some cases also with aromatic substituents placed on the fullerene surface. The next interesting group of complexes created by Indol_20_10 are systems containing FF_3, FF_5, and FF_9 nanocarriers, all of which are characterized by similar affinity values (−8,9 kcal/mol). The increase of yield relative to the reference system in this case reaches a value of 1662.4%. The observed ligand conformations in all cases supports the occurrence of stacking interactions between all aromatic systems of ligand and the surface of considered nanomolecules; denoted distances are placed in the range from 3.27 to 3.8 Å.

Table 7. Values of binding affinity (kcal/mol) of Indol_4_9 molecule relative to functionalized C_{60} fullerene derivatives (FF_X) obtained during docking stage.

Nanostructure Name	ΔG (kcal/mol)				Binding Constant (K_{max})	Difference of K_{max} Relative to C_{60} (%)
	MAX	MIN	AVERAGE	SD		
FF_1	−8.1	−7.6	−7.78	0.12	865722.5	657.9
FF_2	−7.9	−7.6	−7.76	0.07	617698.8	440.8
FF_3	−8.1	−7.8	−7.92	0.10	865722.5	657.9
FF_4	−7.9	−7.7	−7.76	0.07	617698.8	440.8
FF_5	−8.4	−8	−8.14	0.13	1436420.7	1157.5
FF_6	−7.9	−7.5	−7.74	0.10	617698.8	440.8
FF_7	−8	−7.8	−7.91	0.07	731270.0	540.2
FF_8	−7.9	−7.7	−7.79	0.04	617698.8	440.8
FF_9	−8.1	−7.7	−7.94	0.13	865722.5	657.9
FF_10	−7.6	−7.3	−7.47	0.08	372283.6	225.9
FF_11	−8.3	−7.7	−7.92	0.15	1213334.8	962.2
FF_12	−8.7	−8.2	−8.37	0.11	2383332.1	1986.5
FF_13	−8	−7.6	−7.80	0.13	731270.0	540.2
C_{60}	−6.9	−6.8	−6.84	0.05	114226.6	0.0

Table 8. Values of binding affinity (kcal/mol) of Indol_20_10 molecule relative to functionalized C_{60} fullerene derivatives (FF_X) obtained during docking stage.

Nanostructure Name	ΔG (kcal/mol)				Binding Constant (K_{max})	Difference of K_{max} Relative to C_{60} (%)
	MAX	MIN	AVERAGE	SD		
FF_1	−8.60	−8.10	−8.31	0.12	2013184.5	962.2
FF_2	−8.40	−7.80	−8.14	0.19	1436420.7	657.9
FF_3	−8.90	−8.30	−8.56	0.15	3340307.9	1662.4
FF_4	−8.30	−8.00	−8.15	0.09	1213334.8	540.2
FF_5	−8.90	−8.60	−8.71	0.10	3340307.9	1662.4
FF_6	−8.50	−8.10	−8.24	0.10	1700523.4	797.2
FF_7	−8.70	−8.50	−8.63	0.07	2383332.1	1157.5
FF_8	−8.70	−8.30	−8.51	0.11	2383332.1	1157.5
FF_9	−8.90	−8.30	−8.61	0.17	3340307.9	1662.4
FF_10	−8.00	−7.70	−7.89	0.12	731270.0	285.8
FF_11	−9.20	−7.80	−8.31	0.35	5542292.2	2824.3
FF_12	−9.20	−8.60	−8.86	0.19	5542292.2	2824.3
FF_13	−8.60	−8.20	−8.36	0.11	2013184.5	962.2
C_{60}	−7.20	−7.10	−7.14	0.05	189526.5	0.0

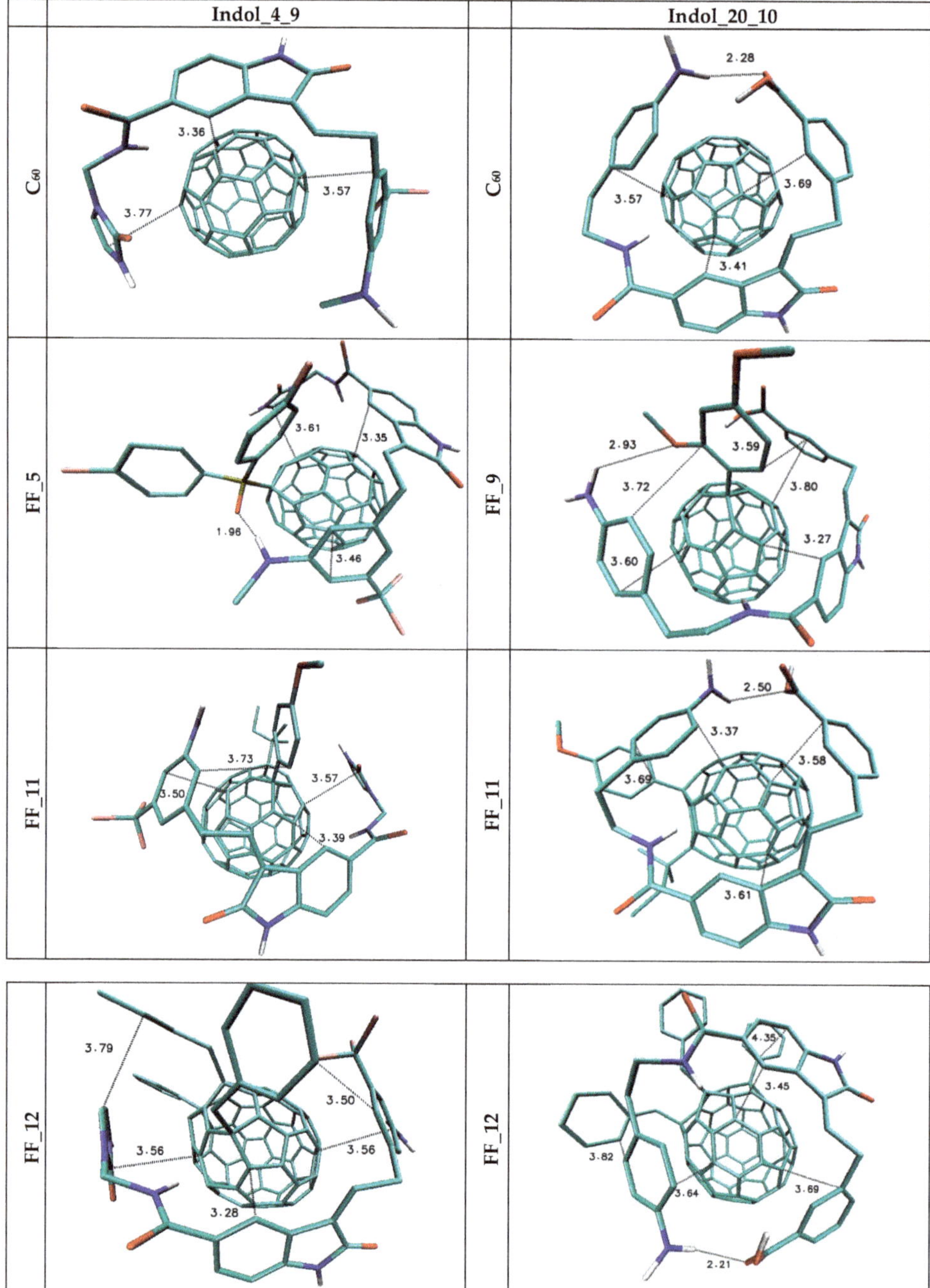

Figure 5. Graphic representation of chosen ligand complexes with functionalized derivatives of C_{60} fullerene characterized by highest binding affinity.

6. Conclusions

The proposed procedure of oxindole derivative preparation allows indication of a group of compounds exhibiting an inhibiting potential towards the CDK2 active site. The realized calculations unambiguously show that both proposed additions to the oxindole core considerably affect the binding affinity towards the active site. The use of similar groups of substituents in both side chains allows a conclusion that modifications made in the R2 position contribute to a much greater increase of binding affinity than modifications in the R1 position. Such an observation clearly shows that this part of the active site created by amino acids LEU83, HIE84, GLN85, ASP86, LYS89, and GLU8 exhibits much more activity in the creation of interactions with ligand molecules. The greatest impact on the values of binding affinity was exerted by substituents, which were heterocyclic or aromatic systems with hydrogen-bond donors and acceptors. Among all the created structures, a significant part was compounds exhibiting not only high binding affinity toward the active site but also an acceptable level of toxicity and a satisfactory ability to penetrate cell membranes expressed by LogP values. The oxindole derivatives chosen based on these factors were evaluated during the molecular dynamics stage and almost all of them confirmed their inhibitory abilities. Two of the analyzed structures, namely Indol_4_9 and Indol_20_10, are characterized by the highest enthalpy values describing their affinities relative to the active site. The energetic aspects correlate also with structural analysis of conformers collected during the molecular dynamics stage. The stable interactions during simulation time, created not only by active groups from molecule core but also for both side chains, together with uniform distribution of RMSD values characterizing these ligands, indicate a significant stability of the considered complexes. Slightly worse but also satisfactory properties exhibit complexes formed by Indol_4_10 and Indol_20_15. Only in the case of Indol_9_15 molecular dynamics analysis was there a total disappearance of interactions created by active groups from side chains identified during the docking stage. This phenomenon is not related to the appearance of new competitive interactions. Such structural observations also confirm a significant decrease of binding enthalpy. The realized calculations show that the best oxindole derivatives exhibit significant affinity towards C_{60} fullerene and functionalized derivatives of this nanomolecule. The three cyclic systems in each of the molecules consist of aromatic or heterocyclic groups which can participate in π–π stacking interactions with fullerene surface. The obtained data shows that the most significant increase of binding affinity is related to the presence of additional aromatic substituents on the fullerene surface as confirmed by values characterizing ligand complexes with FF_12 and FF_11 fullerene. The obtained data shows that one of the most important factors stabilizing such complexes is stacking and hydrophobic interactions; the presence of hydrogen-bond donors or acceptors does not play such an important role in the stabilization of such complexes. The presented data clearly shows that the created complexes of oxindole derivatives and considered nanocarriers exhibit significant potential in the creation of immobilized drugs, and can be used in the development of targeted therapies.

Funding: This research received no external funding.

Acknowledgments: This research was supported by PL-Grid Infrastructure (http://www.plgrid.pl/en).

Conflicts of Interest: The author declares that there are no conflicts of interest.

Appendix A

Table A1. The characteristics of affinities and molecular descriptors of analyzed oxindole derivatives. (ΔG—ligand binding affinity; IC—Inhibition constant; Tox—index of toxicity; LogP—octanol/water coefficient; MW—molecular weight; n HbA—number of hydrogen-bond acceptors; n HbD—number of hydrogen-bond donors.).

Name	SMILES	ΔG (kcal/mol)	IC (nM)	Tox	LogP	MW (g/mol)	n HbA	n HbD
indol_1_10	c12c(cc(cc1)C(=O)NCc1nonc1N)/C(=C/Cc1cc(ccc1)C(=O)O)/C(=O)N2	−11	8.65	0.8	2.056	419.4	10	5
indol_1_11	c12c(cc(cc1)C(=O)NCc1nonc1N)/C(=C/Cc1ccnc(n1)N)/C(=O)N2	−10.66	15.35	0.93	0.074	392.38	11	6
indol_1_12	c12c(cc(cc1)C(=O)NCc1nonc1N)/C(=C/CNc1ccc(cc1)Br)/C(=O)N2	−10.48	20.80	0.9	2.85	469.3	9	5
indol_1_13	c12c(cc(cc1)C(=O)NCc1nonc1N)/C(=C/CC1CCNCC1)/C(=O)N2	−10.3	28.18	0.86	1.692	382.42	9	5
indol_1_15	c12c(cc(cc1)C(=O)NCc1nonc1N)/C(=C/CCc1ccc(cc1)S(=O)(=O)N)/C(=O)N2	−10.04	43.71	0.84	0.416	468.5	11	6
indol_1_19	c12c(cc(cc1)C(=O)NCc1nonc1N)/C(=C/CNC(=O)c1ccccc1)/C(=O)N2	−9.56	98.27	0.75	1.136	418.41	10	5
indol_1_3	c12c(cc(cc1)C(=O)NCc1nonc1N)/C(=C/Cc1nc[nH]c1)/C(=O)N2	−10.00	46.76	0.94	0.465	365.35	10	5
indol_1_4	c12c(cc(cc1)C(=O)NCc1nonc1N)/C(=C/Cn1c(=O)[nH]cc1)/C(=O)N2	−10.1	39.50	0.97	−0.015	381.35	11	5
indol_1_5	c12c(cc(cc1)C(=O)NCc1nonc1N)/C(=C/CC1=CC(=NC1=O)NC)/C(=O)N2	−9.62	88.81	0.92	−1.075	407.39	11	5
indol_1_6	c12c(cc(cc1)C(=O)NCc1nonc1N)/C(=C/Cc1cnc[nH]1)/C(=O)N2	−10.1	39.50	0.94	0.227	365.35	10	5
indol_1_9	c12c(cc(cc1)C(=O)NCc1nonc1N)/C(=C/Cc1cc(cc(c1)NC)C(F)(F)F)/C(=O)N2	−11.08	7.56	0.89	3.527	472.43	9	5
indol_4_1	c12c(cc(cc1)C(=O)NCn1c(=O)[nH]cc1)/C(=C/CCc1nonc1N)/C(=O)N2	−10.2	33.37	0.97	0.314	395.38	11	5
indol_4_10	c12c(cc(cc1)C(=O)NCn1c(=O)[nH]cc1)/C(=C/Cc1cc(ccc1)C(=O)O)/C(=O)N2	−11.1	7.30	0.07	2.043	418.41	9	4
indol_4_11	c12c(cc(cc1)C(=O)NCn1c(=O)[nH]cc1)/C(=C/Cc1ccnc(n1)N)/C(=O)N2	−10.7	14.35	0.88	−0.012	391.39	10	5
indol_4_12	c12c(cc(cc1)C(=O)NCn1c(=O)[nH]cc1)/C(=C/CNc1ccc(cc1)Br)/C(=O)N2	−10.4	23.81	0.35	2.801	468.31	8	4
indol_4_13	c12c(cc(cc1)C(=O)NCn1c(=O)[nH]cc1)/C(=C/CC1CCNCC1)/C(=O)N2	−10.4	23.81	0.17	1.408	381.44	8	4
indol_4_15	c12c(cc(cc1)C(=O)NCn1c(=O)[nH]cc1)/C(=C/CCc1ccc(cc1)S(=O)(=O)N)/C(=O)N2	−10.72	13.87	0.01	0.445	467.51	10	5
indol_4_19	c12c(cc(cc1)C(=O)NCn1c(=O)[nH]cc1)/C(=C/CNC(=O)c1ccccc1)/C(=O)N2	−9.28	157.64	0	0.855	417.43	9	4
indol_4_3	c12c(cc(cc1)C(=O)NCn1c(=O)[nH]cc1)/C(=C/Cc1nc[nH]c1)/C(=O)N2	−10	46.76	0.76	0.405	364.37	9	4
indol_4_5	c12c(cc(cc1)C(=O)NCn1c(=O)[nH]cc1)/C(=C/CC1=CC(=NC1=O)NC)/C(=O)N2	−9.66	83.01	0.43	−1.317	406.4	10	4
indol_4_6	c12c(cc(cc1)C(=O)NCn1c(=O)[nH]cc1)/C(=C/Cc1cnc[nH]1)/C(=O)N2	−10.02	45.21	0.76	0.031	364.37	9	4
indol_4_9	c12c(cc(cc1)C(=O)NCn1c(=O)[nH]cc1)/C(=C/Cc1cc(cc(c1)NC)C(F)(F)F)/C(=O)N2	−11	8.65	0.32	3.579	471.44	8	4
indol_9_1	c12c(cc(cc1)C(=O)NCc1cc(cc(c1)NC)C(F)(F)F)/C(=C/CCc1nonc1N)/C(=O)N2	−9.6	91.86	0.88	3.562	486.45	9	5
indol_9_10	c12c(cc(cc1)C(=O)NCc1cc(cc(c1)NC)C(F)(F)F)/C(=C/Cc1cc(ccc1)C(=O)O)/C(=O)N2	−10.54	18.80	0.39	5.116	509.48	7	4
indol_9_11	c12c(cc(cc1)C(=O)NCc1cc(cc(c1)NC)C(F)(F)F)/C(=C/Cc1ccnc(n1)N)/C(=O)N2	−10.24	31.19	0.69	3.281	482.47	8	5
indol_9_12	c12c(cc(cc1)C(=O)NCc1cc(cc(c1)NC)C(F)(F)F)/C(=C/CNc1ccc(cc1)Br)/C(=O)N2	−9.72	75.01	0.59	5.9	559.39	6	4
indol_9_13	c12c(cc(cc1)C(=O)NCc1cc(cc(c1)NC)C(F)(F)F)/C(=C/CC1CCNCC1)/C(=O)N2	−10.22	32.26	0.45	4.809	472.51	6	4

Table A1. *Cont.*

Name	SMILES	ΔG (kcal/mol)	IC (nM)	Tox	LogP	MW (g/mol)	n HbA	n HbD
indol_9_15	c12c(cc(cc1)C(=O)NCc1cc(cc(c1)NC)C(F)(F)F)/C(=C/CCc1ccc(cc1)S(=O)(=O)N)/C(=O)N2	−10.58	17.57	0.19	3.644	558.58	8	5
indol_9_19	c12c(cc(cc1)C(=O)NCc1cc(cc(c1)NC)C(F)(F)F)/C(=C/CNC(=O)c1ccccc1)/C(=O)N2	−9.72	75.01	0.11	4.076	508.5	7	4
indol_9_3	c12c(cc(cc1)C(=O)NCc1cc(cc(c1)NC)C(F)(F)F)/C(=C/Cc1nc[nH]c1)/C(=O)N2	−9.9	55.36	0.66	3.694	455.44	7	4
indol_9_4	c12c(cc(cc1)C(=O)NCc1cc(cc(c1)NC)C(F)(F)F)/C(=C/Cn1c(=O)[nH]cc1)/C(=O)N2	−10.08	40.86	0.24	2.923	471.44	8	4
indol_9_5	c12c(cc(cc1)C(=O)NCc1cc(cc(c1)NC)C(F)(F)F)/C(=C/CC1=CC(=NC1=O)NC)/C(=O)N2	−9.86	59.23	0.36	2.07	497.48	8	4
indol_9_6	c12c(cc(cc1)C(=O)NCc1cc(cc(c1)NC)C(F)(F)F)/C(=C/Cc1cnc[nH]1)/C(=O)N2	−9.92	53.52	0.67	3.455	455.44	7	4
indol_10_1	c12c(cc(cc1)C(=O)NCc1cc(ccc1)C(=O)O)/C(=C/CCc1nonc1N)/C(=O)N2	−9.54	101.65	0.79	2.235	433.42	10	5
indol_10_11	c12c(cc(cc1)C(=O)NCc1cc(ccc1)C(=O)O)/C(=C/Cc1ccnc(n1)N)/C(=O)N2	−10.7	14.35	0.4	1.66	429.44	9	5
indol_10_12	c12c(cc(cc1)C(=O)NCc1cc(ccc1)C(=O)O)/C(=C/CNc1ccc(cc1)Br)/C(=O)N2	−10.16	35.70	0.51	4.497	506.36	7	4
indol_10_13	c12c(cc(cc1)C(=O)NCc1cc(ccc1)C(=O)O)/C(=C/CC1CCNCC1)/C(=O)N2	−10.36	25.47	0.28	3.377	419.48	7	4
indol_10_15	c12c(cc(cc1)C(=O)NCc1cc(ccc1)C(=O)O)/C(=C/CCc1ccc(cc1)S(=O)(=O)N)/C(=O)N2	−10.58	17.57	0.45	2.318	505.55	9	5
indol_10_19	c12c(cc(cc1)C(=O)NCc1cc(ccc1)C(=O)O)/C(=C/CNC(=O)c1ccccc1)/C(=O)N2	−9.3	152.41	0.03	2.936	455.47	8	4
indol_10_3	c12c(cc(cc1)C(=O)NCc1cc(ccc1)C(=O)O)/C(=C/Cc1nc[nH]c1)/C(=O)N2	−10.18	34.51	0.46	2.176	402.41	8	4
indol_10_4	c12c(cc(cc1)C(=O)NCc1cc(ccc1)C(=O)O)/C(=C/Cn1c(=O)[nH]cc1)/C(=O)N2	−10.44	22.25	0.08	1.582	418.41	9	4
indol_10_5	c12c(cc(cc1)C(=O)NCc1cc(ccc1)C(=O)O)/C(=C/CC1=CC(=NC1=O)NC)/C(=O)N2	−9.98	48.37	0.15	0.749	444.45	9	4
indol_10_6	c12c(cc(cc1)C(=O)NCc1cc(ccc1)C(=O)O)/C(=C/Cc1cnc[nH]1)/C(=O)N2	−10.16	35.70	0.45	1.903	402.41	8	4
indol_10_9	c12c(cc(cc1)C(=O)NCc1cc(ccc1)C(=O)O)/C(=C/Cc1cc(cc(c1)NC)C(F)(F)F)/C(=O)N2	−10.88	10.59	0.46	5.274	509.48	7	4
indol_15_1	c12c(cc(cc1)C(=O)NCc1ccc(cc1)S(=O)(=O)N)/C(=C/CCc1nonc1N)/C(=O)N2	−10.08	40.86	0.87	−1.335	468.5	11	6
indol_15_10	c12c(cc(cc1)C(=O)NCc1ccc(cc1)S(=O)(=O)N)/C(=C/Cc1cc(ccc1)C(=O)O)/C(=O)N2	−10.78	12.54	0.49	0.998	491.52	9	5
indol_15_11	c12c(cc(cc1)C(=O)NCc1ccc(cc1)S(=O)(=O)N)/C(=C/Cc1ccnc(n1)N)/C(=O)N2	−10.48	20.80	0.24	−0.85	464.51	10	6
indol_15_12	c12c(cc(cc1)C(=O)NCc1ccc(cc1)S(=O)(=O)N)/C(=C/CNc1ccc(cc1)Br)/C(=O)N2	−10.4	23.81	0.3	1.952	541.43	8	5
indol_15_13	c12c(cc(cc1)C(=O)NCc1ccc(cc1)S(=O)(=O)N)/C(=C/CC1CCNCC1)/C(=O)N2	−10.08	40.86	0.14	0.662	454.55	8	5
indol_15_19	c12c(cc(cc1)C(=O)NCc1ccc(cc1)S(=O)(=O)N)/C(=C/CNC(=O)c1ccccc1)/C(=O)N2	−10.28	29.15	0.01	0.095	490.54	9	5
indol_15_3	c12c(cc(cc1)C(=O)NCc1ccc(cc1)S(=O)(=O)N)/C(=C/Cc1nc[nH]c1)/C(=O)N2	−10	46.76	0.29	−0.86	437.48	9	5
indol_15_4	c12c(cc(cc1)C(=O)NCc1ccc(cc1)S(=O)(=O)N)/C(=C/Cn1c(=O)[nH]cc1)/C(=O)N2	−10.2	33.37	0.01	−1.221	453.48	10	5
indol_15_5	c12c(cc(cc1)C(=O)NCc1ccc(cc1)S(=O)(=O)N)/C(=C/CC1=CC(=NC1=O)NC)/C(=O)N2	−9.1	213.61	0.02	−2.226	479.52	10	5
indol_15_6	c12c(cc(cc1)C(=O)NCc1ccc(cc1)S(=O)(=O)N)/C(=C/Cc1cnc[nH]1)/C(=O)N2	−10.02	45.21	0.27	−0.753	437.48	9	5
indol_15_9	c12c(cc(cc1)C(=O)NCc1ccc(cc1)S(=O)(=O)N)/C(=C/Cc1cc(cc(c1)NC)C(F)(F)F)/C(=O)N2	−9.36	137.73	0.25	2.576	544.55	8	5
indol_18_1	c12c(cc(cc1)C(=O)N(C(=O)Nc1ccccc1)C)/C(=C/CCc1nonc1N)/C(=O)N2	−9.88	57.26	0.94	0.754	432.44	10	4
indol_18_10	c12c(cc(cc1)C(=O)N(C(=O)Nc1ccccc1)C)/C(=C/Cc1cc(ccc1)C(=O)O)/C(=O)N2	−10.88	10.59	0.03	2.979	455.47	8	3
indol_18_11	c12c(cc(cc1)C(=O)N(C(=O)Nc1ccccc1)C)/C(=C/Cc1ccnc(n1)N)/C(=O)N2	−10.52	19.44	0.7	1.089	428.45	9	4
indol_18_12	c12c(cc(cc1)C(=O)N(C(=O)Nc1ccccc1)C)/C(=C/CNc1ccc(cc1)Br)/C(=O)N2	−10.54	18.80	0.28	3.852	505.37	7	3
indol_18_13	c12c(cc(cc1)C(=O)N(C(=O)Nc1ccccc1)C)/C(=C/CC1CCNCC1)/C(=O)N2	−9.62	88.81	0.11	2.613	418.5	7	3

Table A1. *Cont.*

Name	SMILES	ΔG (kcal/mol)	IC (nM)	Tox	LogP	MW (g/mol)	n HbA	n HbD
indol_18_15	c12c(cc(cc1)C(=O)N(C(=O)Nc1ccccc1)C)/C(=C/CCc1ccc(cc1)S(=O)(=O)N)/C(=O)N2	−10.44	22.25	0	1.649	504.57	9	4
indol_18_19	c12c(cc(cc1)C(=O)N(C(=O)Nc1ccccc1)C)/C(=C/CNC(=O)c1ccccc1)/C(=O)N2	−9.16	193.03	0	1.9	454.49	8	3
indol_18_3	c12c(cc(cc1)C(=O)N(C(=O)Nc1ccccc1)C)/C(=C/Cc1nc[nH]c1)/C(=O)N2	−9.86	59.23	0.57	1.49	401.43	8	3
indol_18_4	c12c(cc(cc1)C(=O)N(C(=O)Nc1ccccc1)C)/C(=C/Cn1c(=O)[nH]cc1)/C(=O)N2	−10.06	42.26	0.06	0.688	417.43	9	3
indol_18_5	c12c(cc(cc1)C(=O)N(C(=O)Nc1ccccc1)C)/C(=C/CC1=CC(=NC1=O)NC)/C(=O)N2	−9.76	70.12	0.12	0.137	443.46	9	3
indol_18_6	c12c(cc(cc1)C(=O)N(C(=O)Nc1ccccc1)C)/C(=C/Cc1cnc[nH]1)/C(=O)N2	−10.07	41.20	0.56	1.264	401.43	8	3
indol_18_9	c12c(cc(cc1)C(=O)N(C(=O)Nc1ccccc1)C)/C(=C/Cc1cc(cc(c1)NC)C(F)(F)F)/C(=O)N2	−10.34	26.34	0.21	4.55	508.5	7	3
indol_20_1	c12c(cc(cc1)C(=O)NCCc1ccc(cc1)N)/C(=C/Cc1cc(ccc1)C(=O)O)/C(=O)N2	−10.4	23.81	0.39	1.261	441.49	7	5
indol_20_10	c12c(cc(cc1)C(=O)NCCc1ccc(cc1)N)/C(=C/Cc1cc(ccc1)C(=O)O)/C(=O)N2	−11.2	6.17	0.39	3.606	441.49	7	5
indol_20_11	c12c(cc(cc1)C(=O)NCCc1ccc(cc1)N)/C(=C/Cc1ccnc(n1)N)/C(=O)N2	−10.7	14.35	0.46	1.814	414.47	8	6
indol_20_12	c12c(cc(cc1)C(=O)NCCc1ccc(cc1)N)/C(=C/CNc1ccc(cc1)Br)/C(=O)N2	−10.4	23.81	0.54	4.528	491.39	6	5
indol_20_13	c12c(cc(cc1)C(=O)NCCc1ccc(cc1)N)/C(=C/CC1CCNCC1)/C(=O)N2	−10.6	16.99	0.4	3.151	404.51	6	5
indol_20_15	c12c(cc(cc1)C(=O)NCCc1ccc(cc1)N)/C(=C/CCc1ccc(cc1)S(=O)(=O)N)/C(=O)N2	−10.8	12.12	0.19	2.184	490.58	8	6
indol_20_19	c12c(cc(cc1)C(=O)NCCc1ccc(cc1)N)/C(=C/CNC(=O)c1ccccc1)/C(=O)N2	−9.9	55.36	0.1	2.488	440.5	7	5
indol_20_3	c12c(cc(cc1)C(=O)NCCc1ccc(cc1)N)/C(=C/Cc1nc[nH]c1)/C(=O)N2	−10.3	28.18	0.49	2.065	387.44	7	5
indol_20_4	c12c(cc(cc1)C(=O)NCCc1ccc(cc1)N)/C(=C/Cn1c(=O)[nH]cc1)/C(=O)N2	−10.4	23.81	0.11	1.461	403.44	8	5
indol_20_5	c12c(cc(cc1)C(=O)NCCc1ccc(cc1)N)/C(=C/CC1=CC(=NC1=O)NC)/C(=O)N2	−9.88	57.26	0.16	0.231	429.48	8	5
indol_20_6	c12c(cc(cc1)C(=O)NCCc1ccc(cc1)N)/C(=C/Cc1cnc[nH]1)/C(=O)N2	−10.28	29.15	0.49	1.849	387.44	7	5
indol_20_9	c12c(cc(cc1)C(=O)NCCc1ccc(cc1)N)/C(=C/Cc1cc(cc(c1)NC)C(F)(F)F)/C(=O)N2	−10.88	10.59	0.5	5.415	494.52	6	5

Table 2. Description of functionalized C_{60} fullerene derivatives (FF_X) used during docking stage.

FF_1	CID_11332103
	$C_{67}H_{14}F_3O_4P$
	1-(diethoxyphosphorylmethyl)-7-(2,2,2-trifluoroethoxy)(C_{60}-I_h)[5,6]fullerene
FF_2	CID_11468612
	$C_{65}H_{13}O_3P$
	9-(diethoxyphosphorylmethyl)-1H-(C_{60}-I_h)[5,6]fullerene
FF_3	CID_16146387
	$C_{67}H_{16}O_2Si$
	methyl 9-(2-trimethylsilylethyl)(C_{60}-I_h)[5,6]fullerene-1-carboxylate
FF_4	CID_16150529
	$C_{70}H_{20}N_2O_2$
	1-N,1-N,9-N,9-N-tetraethyl(C_{60}-I_h)[5,6]fullerene-1,9-dicarboxamide
FF_5	CID_16156307
	$C_{72}H_9F_2OP$
	9-bis(4-fluorophenyl)phosphoryl-1H-(C_{60}-I_h)[5,6]fullerene
FF_6	CID_53469305
	$C_{64}H_{11}O_3P$
	9-diethoxyphosphoryl-1H-(C_{60}-I_h)[5,6]fullerene
FF_7	CID_71618962
	$C_{68}H_{10}O_2$
	9-(3,5-dimethoxyphenyl)-1H-(C_{60}-I_h)[5,6]fullerene
FF_8	CID_71619055
	$C_{68}H_{10}O_2$
	9-(2,6-dimethoxyphenyl)-1H-(C_{60}-I_h)[5,6]fullerene
FF_9	CID_71619159
	$C_{68}H_{10}O_2$
	9-(2,4-dimethoxyphenyl)-1H-(C_{60}-I_h)[5,6]fullerene
FF_10	CID_101218232
	$C_{63}H_4ClF_3O$
	1-(chloromethyl)-7-(2,2,2-trifluoroethoxy)(C_{60}-I_h)[5,6]fullerene
FF_11	CID_101218236
	$C_{69}H_9Cl_3O$
	1-(4-methoxyphenyl)-7-(1,1,2-trichloroethyl)(C_{60}-I_h)[5,6]fullerene
FF_12	CID_101266715
	$C_{80}H_{22}$
	12,15-dibenzyl-9-phenyl-6,18-dihydro-1H-(C_{60}-I_h)[5,6]fullerene
FF_13	CID_101382121
	$C_{62}F_6$
	1,9-bis(trifluoromethyl)(C_{60}-I_h)[5,6]fullerene

References

1. Besson, A.; Dowdy, S.F. Roberts JM CDK Inhibitors: Cell Cycle Regulators and Beyond. *Dev. Cell* **2008**, *14*, 159–169. [CrossRef] [PubMed]
2. Malumbres, M. Barbacid M Cell cycle, CDKs and cancer: A changing paradigm. *Nat. Rev. Cancer* **2009**, *9*, 153–166. [CrossRef] [PubMed]
3. Morgan, D.O. CYCLIN-DEPENDENT KINASES: Engines, Clocks, and Microprocessors. *Annu. Rev. Cell Dev. Biol.* **1997**, *13*, 261–291. [CrossRef] [PubMed]
4. Malumbres, M.; Barbacid, M. To cycle or not to cycle: A critical decision in cancer. *Nat. Rev. Cancer* **2001**, *1*, 222–231. [CrossRef] [PubMed]
5. Child, E.S.; Hendrychová, T.; McCague, K.; Futreal, A.; Otyepka, M.; Mann, D.J. A cancer-derived mutation in the PSTAIRE helix of cyclin-dependent kinase 2 alters the stability of cyclin binding. *Biochim. Biophys. Acta. Mol. Cell Res.* **2010**, *1803*, 858–864. [CrossRef]
6. Echalier, A.; Hole, A.J.; Lolli, G.; Endicott, J.A.; Noble, M.E. An Inhibitor's-Eye View of the ATP-Binding Site of CDKs in Different Regulatory States. *ACS Chem. Biol.* **2014**, *9*, 1251–1256. [CrossRef] [PubMed]
7. Canavese, M.; Santo, L.; Raje, N. Cyclin dependent kinases in cancer. *Cancer Biol. Ther.* **2012**, *13*, 451–457. [CrossRef]
8. LEE, M.Y.; Liu, Y.W.; Chen, M.H.; Wu, J.Y.; Ho, H.Y.; Wang, Q.F.; Chuang, J.J. Indirubin-3'-monoxime promotes autophagic and apoptotic death in JM1 human acute lymphoblastic leukemia cells and K562 human chronic myelogenous leukemia cells. *Oncol. Rep.* **2013**, *29*, 2072–2078. [CrossRef]
9. Hoessel, R.; Leclerc, S.; Endicott, J.A.; Nobel, M.E.; Lawrie, A.; Tunnah, P.; Niederberger, E. Indirubin the active constituent of a Chinese antileukaemia medicine, inhibits cyclin-dependent kinases. *Nat. Cell Biol.* **1999**, *1*, 60–67. [CrossRef]
10. Shin, E.K.; Kim, J.K. Indirubin derivative E804 inhibits angiogenesis. *BMC Cancer* **2012**, *12*, 164. [CrossRef]
11. Marko, D.; Schätzle, S.; Friedel, A.; Genzlinger, A.; Zankl, H.; Meijer, L.; Eisenbrand, G. Inhibition of cyclin-dependent kinase 1 (CDK1) by indirubin derivatives in human tumour cells. *Br. J. Cancer* **2012**, *84*, 283–289. [CrossRef]
12. Martin, L.; Magnaudeix, A.; Wilson, C.M.; Yardin, C.; Terro, F. The new indirubin derivative inhibitors of glycogen synthase kinase-3, 6-BIDECO and 6-BIMYEO, prevent tau phosphorylation and apoptosis induced by the inhibition of protein phosphatase-2A by okadaic acid in cultured neurons. *J. Neurosci. Res.* **2011**, *89*, 1802–1811. [CrossRef]
13. Leclerc, S.; Garnier, M.; Hoessel, R.; Marko, D.; Bibb, J.A.; Snyder, G.L.; Eisenbrand, G. Indirubins Inhibit Glycogen Synthase Kinase-3beta and CDK5/P25, Two Protein Kinases Involved in Abnormal Tau Phosphorylation in Alzheimer's Disease. A PROPERTY COMMON TO MOST CYCLIN-DEPENDENT KINASE INHIBITORS? *J. Biol. Chem.* **2001**, *276*, 251–260. [CrossRef]
14. Beauchard, A.; Ferandin, Y.; Frère, S.; Lozach, O.; Blairvacq, M.; Meijer, L.; Besson, T. Synthesis of novel 5-substituted indirubins as protein kinases inhibitors. *Bioorg. Med. Chem.* **2006**, *14*, 6434–6443. [CrossRef]
15. Dermatakis, A.; Luk, K.C.; DePinto, W. Synthesis of potent oxindole CDK2 inhibitors. *Bioorg. Med. Chem.* **2003**, *11*, 1873–1881. [CrossRef]
16. Luk, K.C.; Simcox, M.E.; Schutt, A.; Rowan, K.; Thompson, T.; Chen, Y.; Dermatakis, A. A new series of potent oxindole inhibitors of CDK2. *Bioorg. Med. Chem. Lett.* **2004**, *14*, 913–917. [CrossRef]
17. Czeleń Molecular dynamics study on inhibition mechanism of CDK-2 and GSK-3β by CHEMBL272026 molecule. *Struct. Chem.* **2016**, *27*, 1807–1818. [CrossRef]
18. Czeleń, P. Inhibition mechanism of CDK-2 and GSK-3β by a sulfamoylphenyl derivative of indoline—A molecular dynamics study. *J. Mol. Model.* **2017**, *23*, 230. [CrossRef]
19. Czeleń, P.; Szefler, B. The Immobilization of Oxindole Derivatives with Use of Cube Rhombellane Homeomorphs. *Symmetry* **2019**, *11*, 900.
20. Szefler, B.; Czeleń, P. Docking of Cisplatin on Fullerene Derivatives and Some Cube Rhombellane Functionalized Homeomorphs. *Symmetry* **2019**, *11*, 874. [CrossRef]
21. Szefler, B.; Czeleń, P.; Diudea, M.V. Docking of indolizine derivatives on cube rhombellane functionalized homeomorphs. *Stud. Univ. Babes-Bolyai Chem.* **2018**, *63*, 7–18. [CrossRef]

22. Morgen, M.; Bloom, C.; Beyerinck, R.; Bello, A.; Song, W.; Wilkinson, K.; Shamblin, S. Polymeric Nanoparticles for Increased Oral Bioavailability and Rapid Absorption Using Celecoxib as a Model of a Low-Solubility, High-Permeability Drug. *Pharm. Res.* **2012**, *29*, 427–440. [CrossRef]
23. De Jong, W.H.; Borm, P.J.A. Drug delivery and nanoparticles:applications and hazards. *Int. J. Nanomed.* **2008**, *3*, 133–149. [CrossRef]
24. Gao, Z.; Zhang, L.; Sun, Y. Nanotechnology applied to overcome tumor drug resistance. *J. Control. Release* **2012**, *162*, 45–55. [CrossRef]
25. Turov, V.V.; Chehun, V.F.; Barvinchenko, V.N.; Krupskaya, T.V.; Prylutskyy, Y.I.; Scharff, P.; Ritter, U. Low-temperature 1H-NMR spectroscopic study of doxorubicin influence on the hydrated properties of nanosilica modified by DNA. *J. Mater. Sci. Mater. Med.* **2011**, *22*, 525–532. [CrossRef]
26. Szefler, B. Nanotechnology, from quantum mechanical calculations up to drug delivery. *Int. J. Nanomed.* **2018**, *13*, 6143–6176. [CrossRef]
27. Cataldo, F.; Da Ros, T. *Medicinal Chemistry and Pharmacological Potential of Fullerenes and Carbon Nanotubes*; Springer: Amsterdam, The Netherlands, 2008.
28. Panchuk, R.R.; Prylutska, S.V.; Chumak, V.V.; Skorokhyd, N.R.; Lehka, L.V.; Evstigneev, M.P.; Ritter, U. Application of C60 Fullerene-Doxorubicin Complex for Tumor Cell Treatment In Vitro and In Vivo. *J. Biomed. Nanotechnol.* **2015**, *11*, 1139–1152. [CrossRef]
29. Andrievsky, G.; Klochkov, V.; Derevyanchenko, L. Is the C_{60} Fullerene Molecule Toxic? *Fuller. Nanotub. Carbon Nanostruct.* **2005**, *13*, 363–376. [CrossRef]
30. Prylutska, S.; Bilyy, R.; Overchuk, M.; Bychko, A.; Andreichenko, K.; Stoika, R.; Ritter, U. Water-soluble pristine fullerenes C60 increase the specific conductivity and capacity of lipid model membrane and form the channels in cellular plasma membrane. *J. Biomed. Nanotechnol.* **2012**, *8*, 522–527. [CrossRef]
31. Qiao, R.; Roberts, A.P.; Mount, A.S.; Klaine, S.J.; Ke, P.C. Translocation of C60 and Its Derivatives Across a Lipid Bilayer. *Nano Lett.* **2007**, *7*, 614–619. [CrossRef]
32. TURBOMOLE 7.0. Available online: http://www.turbomole.com/.
33. Potemkin, V.; Grishina, M. Principles for 3D/4D QSAR classification of drugs. *Drug. Discov. Today* **2008**, *13*, 952–959. [CrossRef]
34. Potemkin, V.A.; Grishina, M.A. A new paradigm for pattern recognition of drugs. *J. Comput. Aided Mol. Des.* **2008**, *22*, 489–505. [CrossRef]
35. Chemosophia. Available online: http://www.chemosophia.com/ (accessed on 1 March 2017).
36. Eckert, F.; Klamt, A. Fast solvent screening via quantum chemistry: COSMO-RS approach. *AIChE J.* **2002**, *48*, 369–385. [CrossRef]
37. Trott, O.; Olson, A.J. AutoDock Vina: Improving the speed and accuracy of docking with a new scoring function, efficient optimization, and multithreading. *J. Comput. Chem.* **2010**, *31*, 455–461. [CrossRef]
38. PubChem. Available online: https://pubchem.ncbi.nlm.nih.gov/ (accessed on 1 May 2019).
39. Bartashevich, E.V.; Potemkin, V.A.; Grishina, M.A.; Belik, A.V. A Method for Multiconformational Modeling of the Three-Dimensional Shape of a Molecule. *J. Struct. Chem.* **2002**, *43*, 1033–1039. [CrossRef]
40. Rosita, G.; Manuel, C.; Franco, M.; Cinzia, N.; Donatella, F.; Emiliano, L.; Roberta, G. Permethrin and its metabolites affect Cu/Zn superoxide conformation: Fluorescence and in silico evidences. *Mol. Biosyst.* **2015**, *11*, 208–217. [CrossRef]
41. Mangiaterra, G.; Laudadio, E.; Cometti, M. Inhibitors of multidrug efflux pumps of Pseudomonas aeruginosa from natural sources: An in silico high-throughput virtual screening and in vitro validation. *Med. Chem. Res.* **2017**, *26*, 414–430. [CrossRef]
42. Maier, J.A.; Martinez, C.; Kasavajhala, K.; Wickstrom, L.; Hauser, K.E.; Simmerling, C. ff14SB: Improving the Accuracy of Protein Side Chain and Backbone Parameters from ff99SB. *J. Chem. Theory Comput.* **2015**, *11*, 3696–3713. [CrossRef]
43. Bayly, C.I.; Cieplak, P.; Cornell, W.; Kollman, P.A. A well-behaved electrostatic potential based method using charge restraints for deriving atomic charges: The RESP model. *J. Phys. Chem.* **1993**, *97*, 10269–10280. [CrossRef]
44. Adelman, S.A. Generalized Langevin equation approach for atom/solid-surface scattering: General formulation for classical scattering off harmonic solids. *J. Chem. Phys.* **1976**, *64*, 2375. [CrossRef]
45. Humphrey, W.; Dalke, A.; Schulten, K. VMD: Visual molecular dynamics. *J. Mol. Graph.* **1996**, *14*, 33–38. [CrossRef]

46. Miller, B.R.; McGee, T.D.; Swails, J.M. *MMPBSA.py*: An Efficient Program for End-State Free Energy Calculations. *J. Chem. Theory Comput.* **2012**, *8*, 3314–3321. [CrossRef]
47. Case, D.A.; Babin, V.; Berryman, J.T.; Betz, R.M.; Cai, Q.; Cerutti, D.S.; Cheatham, T.E., III; Darden, T.A.; Duke, R.E.; Gohlke, H.; et al. *AMBER 14*; University of California: Oakland, CA, USA, 2014.

Article

Predicting Value of Binding Constants of Organic Ligands to Beta-Cyclodextrin: Application of MARSplines and Descriptors Encoded in SMILES String

Piotr Cysewski and Maciej Przybyłek *

Department of Physical Chemistry, Faculty of Pharmacy, Collegium Medicum of Bydgoszcz, Nicolaus Copernicus University in Toruń, Kurpińskiego 5, 85-950 Bydgoszcz, Poland
* Correspondence: m.przybylek@cm.umk.pl

Received: 29 June 2019; Accepted: 12 July 2019; Published: 15 July 2019

Abstract: The quantitative structure–activity relationship (QSPR) model was formulated to quantify values of the binding constant (lnK) of a series of ligands to beta–cyclodextrin (β-CD). For this purpose, the multivariate adaptive regression splines (MARSplines) methodology was adopted with molecular descriptors derived from the simplified molecular input line entry specification (SMILES) strings. This approach allows discovery of regression equations consisting of new non-linear components (basis functions) being combinations of molecular descriptors. The model was subjected to the standard internal and external validation procedures, which indicated its high predictive power. The appearance of polarity-related descriptors, such as XlogP, confirms the hydrophobic nature of the cyclodextrin cavity. The model can be used for predicting the affinity of new ligands to β-CD. However, a non-standard application was also proposed for classification into Biopharmaceutical Classification System (BCS) drug types. It was found that a single parameter, which is the estimated value of lnK, is sufficient to distinguish highly permeable drugs (BCS class I and II) from low permeable ones (BCS class II and IV). In general, it was found that drugs of the former group exhibit higher affinity to β-CD then the latter group (class III and IV).

Keywords: beta-cyclodextrin; QSPR; binding constant

1. Introduction

Molecular complexes, such as inclusion adducts, clathrates, cocrystals and solvates, have been widely used in many fields, including pharmacy [1–6], agriculture [7], the food industry [1,8,9] and explosives [10,11]. In the past two decades, cyclodextrins (CDs) have been one of the most extensively studied complexation agents, especially as pharmaceutical excipients [12–15]. The CDs' adducts with active pharmaceutical ingredients (APIs) are mainly used for solubility and bioavailability enhancement [12–17], stability improvement [13,18], stomach, skin and eye irritation reduction [13,17,19], and prevention of unpleasant odor and bitter taste [13,20]. Probably the most commonly used compounds in pharmaceutical formulations belonging to this class are alpha- (α-CD), beta- (β-CD), gamma- (γ-CD) cyclodextrins and their analogues such as (2-hydroxypropyl)-beta-cyclodextrin (HP-β-CD), sulfobutylether beta-cyclodextrin (SBE-β-CD) or randomly methylated-β-cyclodextrins (RM-β-CDs) [21]. The main criterion used for distinguishing different CDs (α-CD, β-CD and γ-CD) corresponds to six, seven and eight D-glucopyranose units, respectively. These excipients have been used for all main types of drug delivery systems (oral, nasal, rectal, dermal, ocular, parenteral) [13,21].

Apart from pharmaceutical applications, cyclodextrins are widely used in personal care products and perfumes manufacturing [20,22,23], which relies on their high stability, solubilizing abilities and vapor pressure reduction (fragrances industry). The unique properties of cyclodextrins are related to their specific structural features. These compounds are characterized by a symmetric toroidal shape. Due to the relatively hydrophobic cavity, cyclodextrins play the role of hosts in molecular inclusion complexes. On the other hand, they are hydrophilic on the outside surface, which results in strong interactions with water molecules. This characteristic structure is somewhat similar to biocatalysts [24–26]. An interesting example of the use of such cyclodextrin-based artificial enzymes is asymmetric and stereospecific synthesis (halogenation, hydrohalogenation, oxidation, reduction, photolysis, aldol reactions, hydrogenation, substitution and addition reaction) [25]. Noteworthy, the stereoselectivity of CDs was utilized for separation of racemic mixtures [27,28].

Quantitative structure–activity relationship (QSPR) methodology has been extensively used for evaluating the formation abilities of molecular complexes, and characterizing their properties [4,29–38]. Cyclodextrin binding constant modeling deserves special attention due to its practical importance. In recent years, several interesting approaches have appeared, such as application of molecular docking [38], conductor like screening model for real solvents (COSMO-RS) and quantum chemical-based descriptors [31], and topological indices [32,34,39]. Most of these models are simple regression equations. In general, better accuracy can be achieved when non-linear methods are applied. In our previous work, a novel approach of combining the non-linear MARSplines (multivariate adaptive regression splines) [40] methodology with common molecular descriptors calculated from simplified molecular input line entry specification (SMILES) code was applied for solubility modeling [41,42]. The major advantage of this procedure is its good predictive power and relatively simple model, which is a regression equation of new factors. The aim of this study was to apply a similar methodology for β-CD stability binding modeling.

2. Materials and Methods

The experimental values used for model development and validation were obtained from the datasets published by Suzuki et al. [43] and Mirrahimi et al. [38]. This collection comprises binding constants of 1:1 β-CD complexes with different organic compounds. In case of experimental values of the same compounds from different sources the mean value was taken into account. The list of all data is provided in the Supplementary Materials (Table S1).

2.1. Molecular Descriptors

Currently, a variety of molecular parameters are freely available for potential applications, which ensures that formulated models can be readily applied for predictions of compounds' properties. Here two online tools were used for collecting the set of descriptors, namely ChemDes [44] and the BioCCl module of the BioTriangle platform [45]. The former provides a direct and integrated way of retrieving the sets of descriptors catalogued as Chemopy Descriptors (1135), CDK Descriptors (275), RDKit Descriptors (196), Pybel Descriptors (24), BlueDesc Descriptors (174) and PaDEL Descriptors 1875). The number of potential parameters is provided in parenthesis. All these indices can be calculated on-line [46]. The second source also offers a limited number of descriptors but offers sets suited for intermolecular interactions, which is the key advantage of this software. This is available using the BioTriangle webserver [47]. Since the BioTriangle allows for different geometrical transformations of descriptors calculated for pairs of molecules, the β-CD-ligand pairs descriptors were included.

2.2. Data Pre-Treatment

After completing the datasets of all descriptors, the standard pre-treatment procedure was implemented. It comprised elimination of descriptors not computable for the whole set of ligands and the remaining content across the whole population. Then, highly correlated and low-variance descriptors were also removed. Data curating was undertaken by taking advantage of the Data PreTreatment 1.2 module relying on the variable reduction Wootton, Sergent and Phan-Tan-Luu's (V-WSP) algorithm [48,49]. Then, the dataset was divided into training and test sets using the activity division approach implemented in the Dataset Division 1.2 tool [50,51]. Both programs are written in Java and are freely available [52].

2.3. Model Development Using MARSplines

In this work, a MARSplines [40] methodology was applied as implemented in STATISTICA 12 [53]. This methodology leads to the following general regression formula, where F_i is the regression factor and a_i are the regression parameters:

$$\ln\left(K_{bCDB}^{est}\right) = a_0 + \sum_{i=1}^{n} a_i \cdot F_i \tag{1}$$

The left-hand side of this equation stand for the response variable, which is confronted with experimental values. Here it is defined by the value of the natural logarithm of a binding constant quantifying affinity of a ligand toward β–CD determined experimentally. The MARSplines method is the procedure designated for finding the analytical formula based on descriptors and so-called knots. Such relationships are termed basis functions. The values represent splitting of the set of values into sub-regions treated with alternative mathematical formula. The number of basis functions and factors in the model is controlled at an arbitrary level for balancing between accuracy and complexity of the model. To avoid model overfitting, the final model undergoes inspection of the regression coefficients by removing such factors for which statistical significance is not reached ($p > 0.05$). Additionally, the contribution to the model of each factor is inferred from the values of standardized regression coefficients (β_i). Only such factors are included in the final model for which $|\beta_i| > 0.09$. Furthermore, the model was refined, internally validated and characterized in terms of fitting criteria using QSARINS software [54–56]. As a result of this procedure, the model was simplified by selecting the most important variables using a genetic algorithm (GA).

The simplest factor generated by the MARSplines procedure has a form identical to the classical QSPR approach and is expressed simply as multiplication of descriptor values by a coefficient, whose value is optimized for maximizing correlations between computed and estimated response values. The main improvement, however, comes from accounting for non-linearity by direct inclusion of more complex basis functions combined into factors. Hence, an advantage of the QSPR model formulation using the MARSplines procedure is the benefit of formally being in the multiple linear regression (MLR) format by including non-linear properties of considered datasets. Hence, the golden standard QSPR model development and validation procedures can be directly applied [41,42,57].

3. Results and Discussion

There are two main reasons that justify the efforts of the obtained model building. The first is obviously of substantive nature for deriving a model that is as accurate as possible and characterized by a low cost of applications. Hence, the screening of new potential ligands, or comparing a leading compound of an API and derivatives suggested by a drug design procedure, represent the immediate value of the obtained model. There is also a methodological reason for exploring the landscape of potential application in the chemistry domain of the MARSplines procedure. This is not explored deeply enough bearing in mind its high potential, effectiveness and ease of use.

3.1. Findings

Based on the MARSplines algorithm, the following descriptors were included in the model (Table 1): XLogP and Wlambda2.unity (source: BlueDesc); carbonTypes.8 (source: CDK), MLFER_A, AATS6m, AATS4i and PNSA-3 (source: PADEL); the tensor product of PEOEVSA9 descriptor vectors denoted as PEOEVSA9*PEOEVSA9; and, the vector sum of Chiv1 parameter denoted as Chiv1plusChiv1 (source: BioTriangle). Taking into account the relatively large training set population ($n = 187$), the number of variables seems to be reasonable fulfilling the general rules of acceptable QSPR model complexity, as documented in Table 1 and Figure 1. Internal validation, fitting criteria and external validation parameters, including R^2 (determination coefficient), R_{adj}^2 (adjustment determination coefficient), F (Fisher ratio), SD (standard deviation), MAE (mean absolute error), MAPE (mean absolute percentage error), RMSE (root-mean-square error), PRESS (predicted residual error sum of squares) and K_{xx} (descriptors' global correlation measure) [58,59], suggest that the model is well fitted to the training set and, most importantly, the external test set examples were well predicted. The results of external validation are presented in Figure 1. As one can see, the proposed model is characterized by high determination coefficients. Interestingly, R^2, MAE and MAPE values are even slightly better for the external test set (0.936, 0.44, 9.3%, respectively) than for the training set (0.907, 0.49, 15.4%, respectively). This suggests that the model complexity is optimal. It is worth mentioning that the over-fitting problem should be taken into account when analyzing the quality of QSPR models, especially those that are non-linear. In the case of overly complex models, the training set data are exceptionally well fitted, but the test set prediction quality is far inferior. It is worth mentioning that the MARSplines protocol implemented in the STATISTICA software prevents overfitting by taking advantage from of the generalized cross validation (GCV) algorithm, which reduces the model to be as simple as possible.

Table 1. Multivariate adaptive regression splines (MARSplines) model parameters along with the validation results.

Factor	β_i	a_i	Basis Functions
F0		5.4277	
F1	−0.3991	−1.2679	max(0; 28.1940-Chiv1plusChiv1)
F2	0.5652	1.1090	max(0; XLogP+0.1340)
F3	0.3772	1.3356	max(0; carbonTypes.8)
F4	−0.1559	−1.4658	max(0; 0.5620-MLFER_A)
F5	−0.1613	−1.3482	max(0; Wlambda2.unity-1.2400)
F6	−0.1391	−0.3182	max(0; 1.2400-Wlambda2.unity)
F7	−0.2130	−2.5385	max(0; XLogP+0.1340)·max(0;-15.8078-PNSA-3)
F8	−0.1372	−0.0120	max(0; 66.0412- AATS6m)·max(0; MLFER_A -0.5620)
F9	−0.0977	−0.0003	max(0; PEOEVSA9*PEOEVSA9-988.3780)·max(0; XLogP +0.1340)
F10	−0.1258	−0.0002	max(0; 988.378- PEOEVSA9*PEOEVSA9)·max(0; XLogP +0.1340)
F11	0.0910	0.0257	max(0; Wlambda2.unity -1.2400)·max(0; AATS4i-154.1756)
F12	0.0944	0.2092	max(0; Wlambda2.unity -1.2400)·max(0; 154.1756- AATS4i)

Model statistics: internal validation (MAE_{CV} = 0.51, $RMSE_{CV}$ = 0.65, Q^2_{LOO} = 0.90, Q^2_{LMO} = 0.90, $PRESS_{CV}$ = 99.00), fitting criteria (N = 187, R^2 = 0.91, R_{adj}^2 = 0.91, MAE_{tr} = 0.61, $RMSE_{tr}$ = 0.48, F = 189.45, SD = 0.63, K_{xx} = 0.35) and external validation (training set: MAE = 0.49, MAPE = 15.4%, test set: MAE = 0.44, MAPE = 9.3%).

Some of the parameters used in the model, such as like XLogP and MLFER_A, are quite intuitive and their physical meaning can be easily explained. The appearance of the hydrophilicity measure, namely the group contribution logP parameter (XlogP), confirms the role of the hydrophobic nature of the cyclodextrin cavity, while MLFER_A is the Abraham solubility parameter expressing the acidity. The role of polarity in β–CD molecular complexes formation was emphasized by PNSA-3 (charged partial surface area index [60]) and BioTriangle interaction descriptor PEOEVSA9*PEOEVSA9. This latter feature was calculated based on the MOE-type parameter involving the contributions of surface area and partial charge [61]. Another feature calculated using the BioTriangle platform, namely Chiv1plusChiv1, is associated with the Chiv1 descriptor belonging to the atomic valence connectivity indices class [62,63]. Of note, these descriptors were widely used in solving quite similar QSAR problems associated with target-ligand binding [64–68]. In the MARSplines model, there was also one topological descriptor characterizing carbon type (carbonTypes.8) [69] and the appearance of two autocorrelation indices, AATS6m and AATS4i [69]. Autocorrelation descriptors are probably one of the most extensively used quantitative structure–activity relationship/quantitative structure property relationship (QSAR/QSPR) descriptors Although the physical meaning of these parameters is not straightforward, our previous studies showed that this broad class of descriptors was found to be useful in the modelling of the affinity of compounds in the solid state [29,57].

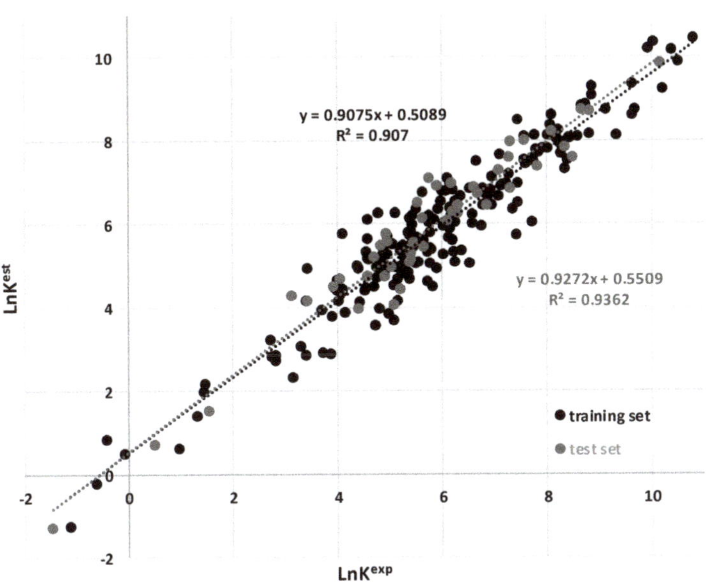

Figure 1. The relationship between experimental and calculated values of binding constants.

3.2. Comparison to Existing Models

Comparison of the determination coefficient calculated for the obtained model with two regression models reported in recent years is presented in Table 2. Although these models were generated using similar datasets, it should be taken into account that depending on the validation procedure, different results can be obtained. Nevertheless, the correlation coefficients are lower or approximately equal to the MARSplines model. This suggests that the proposed approach is a good alternative for β-CD calculation. The major advantage of calculating molecular descriptors from the SMILES code is low computational cost. However, the proposed QSPR model has some limitations associated with ignoring the geometrical features of molecular complexes, such as conformation and solvation effects. These effects can be included using optimized 3D structures. Furthermore, it should be taken into account that in some cases, the stoichiometry of β-CD complexes is not 1:1 [70–72]. In such cases, molecular modelling methods such as molecular-dynamics docking or quantum-chemical binding constant calculations are more appropriate than the proposed approach.

Table 2. Comparison of determination coefficients of beta-cyclodextrin (β-CD) binding constants.

Model Description	R^2		Source
	Training Set	Test Set	
MARSplines	0.91	0.94	This work
Molecular docking-based descriptors	0.83	0.83	[37]
Monte Carlo optimised topological descriptors	0.92	0.93	[38]

3.3. Exemplary Model Applications

The obvious application of the model provided by Equation (1) and Table 1 relates to its predictive power. Hence, it is possible to anticipate, before actual measurement, the probable affinity of the considered API toward β-CD. There is, of course, a limitation due to the applicability domain. For example, there are no organic and metalo-organic salts in the model. Hence, it is very unlikely that the model helps in situations where drugs are prepared in such forms. However, many drugs are, in principle, treatable by the model and at least the rational selection of the candidates for experimental measurements can be advised.

It is also possible to suggest alternative, less obvious applications of the formulated MARSplines model. For example, in the Biopharmaceutical Classification System (BCS) it is assumed that two measures such as solubility and permeability can be used for grouping drugs in respect of their bioavailability. In Table 3 this classification is shown [73]. Of note, cyclodextrins and their solubilizing abilities have been discussed in the context of BCS classification [74,75].

Table 3. Biopharmaceutical Classification System (BCS) [73] using solubility and permeability as qualitative criterions.

	High Solubility	**Low Solubility**
High permeability	Class I This class comprise compounds characterized by good absorption profiles.	Class II The bioavailability is directly related to the dissolution behavior.
Low permeability	Class III The active pharmaceutical ingredient (API) is soluble, however absorption profile is dependent on limited permeation behavior.	Class IV The API is characterized by very low bioavailability.

Hence, for proper bioavailability assessment, both water solubility and permeability must be known. It is interesting to see if there is any correlation between the BCS class of a given drug and its estimated affinity toward β-CD. For this purpose, information about the BCS classification was collected for 300+ drugs [76]. Those that are found to be outside of the applicability domain were excluded from the analysis. For the remaining drugs, the values of the molecular descriptors were collected. This, in turn, allowed for application of the MARSplines model and prediction of lnK values. The obtained results are presented in Figure 2 and Table S2. As can be seen from Figure 2, those APIs exhibiting good permeability (Class I and II) are characterized by higher affinity to cyclodextrin. This is understandable since the cyclodextrin cavity is rather hydrophobic, like for lipid biological barriers. The most important message coming from Figure 2 is that Class I and II have very similar distributions to each other and, at the same time, are distinct from Class III and IV. Indeed, application of a statistical non-parametrical test revealed that the medians are statistically the same ($p = 0.27$) for Class I and II but either combination with remaining classes reached statistical significance ($p < 0.001$). Similarly, the analysis of Classes III and IV versus the other two classes consistently confirms that low permeability can be distinguished from high values by predicted drug affinity to β-CD. In order to turn this qualitative conclusion into a practically useful formula, a second MARSplines model was formulated. However, the target of the modeling this time was the classification into low and high permeability cases. Hence, only one quantitative parameter was used for classification model formulation, namely, computed values of lnK. As a dependent value, the binary flag for permeability was declared. The obtained formulae are provided below:

$$ClassA = 0.1655 + 0.2510 \times \max(0;\ LnK - 4.8148) + 0.0734 \times \max(0;\ 4.8148 - LnK) \\ - 0.2455 \times \max(0;\ LnK - 8.0157) \tag{2}$$

$$ClassB = 0.8345 - 0.2510 \times \max(0;\ LnK - 4.8148) - 0.0734 \times \max(0;\ 4.8148 - LnK) \\ + 0.2455 \times \max(0;\ LnK - 8.0157) \tag{3}$$

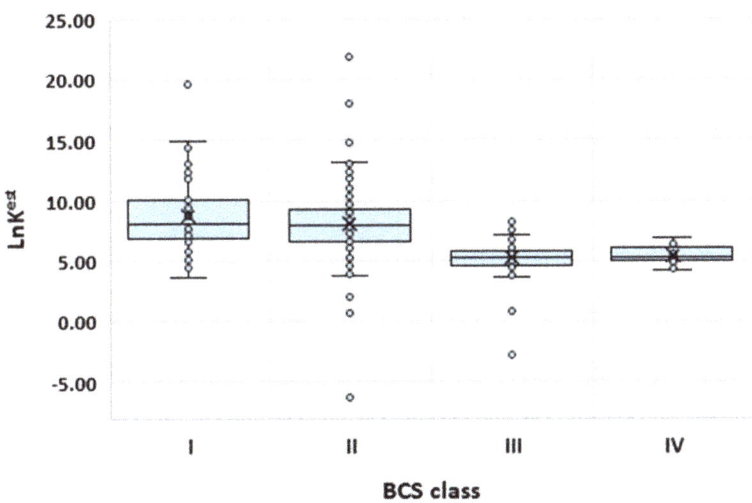

Figure 2. The distributions of affinity of APIs to β-CD as a function of BCS classification.

If the value of the first equation for dependent variable ClassA is higher compared to the value provided by the second equation, then high permeability is predicted. This means that the analyzed drug belongs to Class I or II of the BCS. On the contrary situation, when ClassA < ClassB, then low permeability is predicted by the model and, consequently, the given drug should belong to Class III or IV of the BCS. It is interesting to note that such a simple model has quite an acceptable predictive power. Proper qualification of high permeability occurred in 88% of cases, with only 12% of misclassified drugs. The low permeability was classified with slightly lower precision of 73%, with 27% of failure. These observations indicate the potential applicability of binding constants for evaluating permeability.

4. Conclusions

Compounds exhibiting high symmetry, such as fullerenes or nanotubes, have been used in various branches of medicine and pharmacy, including drug delivery [77–79]. This also applies to cyclodextrins, which due to their specific shape features, have been widely used to increase API solubility. In this work, the QSPR model of the binding constant of different compounds to β-CD was developed based on the MARSplines methodology, and molecular descriptors were derived from the SMILES code. The internal and external validation indicated good accuracy of the model. The appearance of polarity-related descriptors, such as XlogP, indicated the hydrophobic nature of the cyclodextrin cavity, which is consistent with the nature of cyclodextrins. It is well known that the hydrophilicity/hydrophobicity of a drug can be used for evaluation of the drugs' permeability. Therefore, the model was used for predicting affinity to β-CD of exemplary compounds belonging to different BCS classes. As was established, APIs exhibiting high permeability (I and II BCS Class) are generally characterized by higher lnK values than compounds revealing low permeability (class III and IV). This shows that β-CD complexation seems to offer an alternative for complex and expensive experimental permeability modeling studies.

Supplementary Materials: The following are available online at http://www.mdpi.com/2073-8994/11/7/922/s1, Table S1: Experimental and calculated LnK values, Table S2: LnK values predicted for compounds belonging to different classes according to Biopharmaceutical Classification System (BCS).

Author Contributions: Both authors contributed equally to the manuscript.

Funding: This research received no external funding.

Conflicts of Interest: The authors declare no conflict of interest.

References

1. Wang, N.; Xie, C.; Hao, H.; Lu, H.; Lou, Y.; Su, W.; Guo, N. Cocrystal and its Application in the Field of Active Pharmaceutical Ingredients and Food Ingredients. *Curr. Pharm. Des.* **2018**, *24*, 2339–2348. [CrossRef]
2. Korotkova, E.I.; Kratochvíl, B. Pharmaceutical Cocrystals. *Procedia Chem.* **2014**, *10*, 473–476. [CrossRef]
3. Wang, Q.; Xue, J.; Hong, Z.; Du, Y. Pharmaceutical Cocrystal Formation of Pyrazinamide with 3-Hydroxybenzoic Acid: A Terahertz and Raman Vibrational Spectroscopies Study. *Molecules* **2019**, *24*, 488. [CrossRef] [PubMed]
4. Cysewski, P.; Przybyłek, M. Selection of effective cocrystals former for dissolution rate improvement of active pharmaceutical ingredients based on lipoaffinity index. *Eur. J. Pharm. Sci.* **2017**, *107*, 87–96. [CrossRef] [PubMed]
5. Przybyłek, M.; Ziółkowska, D.; Mroczyńska, K.; Cysewski, P. Applicability of Phenolic Acids as Effective Enhancers of Cocrystal Solubility of Methylxanthines. *Cryst. Growth Des.* **2017**, *17*, 2186–2193. [CrossRef]
6. Sinha, A.S.; Maguire, A.R.; Lawrence, S.E. Cocrystallization of nutraceuticals. *Cryst. Growth Des.* **2015**, *15*, 984–1009. [CrossRef]
7. Yang, R.; Xiao, C.F.; Guo, Y.F.; Ye, M.; Lin, J. Inclusion complexes of GA 3 and the plant growth regulation activities. *Mater. Sci. Eng. C* **2018**, *91*, 475–485. [CrossRef]
8. Koontz, J.L.; Marcy, J.E.; Barbeau, W.E.; Duncan, S.E. Stability of Natamycin and Its Cyclodextrin Inclusion Complexes in Aqueous Solution. *J. Agric. Food Chem.* **2003**, *51*, 7111–7114. [CrossRef]
9. Martina, K.; Binello, A.; Lawson, D.; Jicsinszky, L.; Cravotto, G. Recent Applications of Cyclodextrins as Food Additives and in Food Processing. *Curr. Nutr. Food Sci.* **2013**, *9*, 167–179. [CrossRef]
10. Guo, C.; Zhang, H.; Wang, X.; Xu, J.; Liu, Y.; Liu, X.; Huang, H.; Sun, J. Crystal structure and explosive performance of a new CL-20/caprolactam cocrystal. *J. Mol. Struct.* **2013**, *1048*, 267–273. [CrossRef]
11. Shen, J.P.; Duan, X.H.; Luo, Q.P.; Zhou, Y.; Bao, Q.; Ma, Y.J.; Pei, C.H. Preparation and characterization of a novel cocrystal explosive. *Cryst. Growth Des.* **2011**, *11*, 1759–1765. [CrossRef]
12. Loftsson, T.; Brewster, M.E. Pharmaceutical applications of cyclodextrins. 1. Drug solubilization and stabilization. *J. Pharm. Sci.* **1996**, *85*, 1017–1025. [CrossRef]
13. Tiwari, G.; Tiwari, R.; Rai, A. Cyclodextrins in delivery systems: Applications. *J. Pharm. Bioallied Sci.* **2010**, *2*, 72. [CrossRef] [PubMed]
14. Archontaki, H.A.; Vertzoni, M.V.; Athanassiou-Malaki, M.H. Study on the inclusion complexes of bromazepam with β- and β-hydroxypropyl-cyclodextrins. *J. Pharm. Biomed. Anal.* **2002**, *28*, 761–769. [CrossRef]
15. de Miranda, J.C.; Martins, T.E.A.; Veiga, F.; Ferraz, H.G. Cyclodextrins and ternary complexes: Technology to improve solubility of poorly soluble drugs. *Brazilian J. Pharm. Sci.* **2011**, *47*, 665–681. [CrossRef]
16. Arima, H.; Yunomae, K.; Miyake, K.; Irie, T.; Hirayama, F.; Uekama, K. Comparative studies of the enhancing effects of cyclodextrins on the solubility and oral bioavailability of tacrolimus in rats. *J. Pharm. Sci.* **2001**, *90*, 690–701. [CrossRef] [PubMed]
17. Rasheed, A.; Kumar C.K., A.; Sravanthi, V.V.N.S.S. Cyclodextrins as drug carrier molecule: A review. *Sci. Pharm.* **2008**, *76*, 567–598. [CrossRef]
18. Arima, H.; Miyaji, T.; Irie, T.; Hirayama, F.; Uekama, K. Enhancing effect of hydroxypropyl-β-cyclodextrin on cutaneous penetration and activation of ethyl 4-biphenylyl acetate in hairless mouse skin. *Eur. J. Pharm. Sci.* **1998**, *6*, 53–59. [CrossRef]
19. Shimpi, S.; Chauhan, B.; Shimpi, P. Cyclodextrins: application in different routes of drug administration. *Acta Pharm.* **2005**, *55*, 139–156. [PubMed]
20. Sharma, N.; Baldi, A. Exploring versatile applications of cyclodextrins: An overview. *Drug Deliv.* **2016**, *23*, 739–757. [CrossRef]
21. European Medicines Agencs. *Cyclodextrins Used as Excipients Report*; European Medicines Agencs: Amsterdam, The Netherlands, 2017.
22. Numanoğlu, U.; Şen, T.; Tarimci, N.; Kartal, M.; Koo, O.M.Y.; Önyüksel, H. Use of cyclodextrins as a cosmetic delivery system for fragrance materials: Linalool and benzyl acetate. *AAPS PharmSciTech* **2008**, *8*, 34–42. [CrossRef] [PubMed]

23. Buschmann, H. Eckhard Schollmeyer Applications of cyclodextrins in cosmetic products: A review. *J. Cosmet. Sci.* **2002**, *53*, 185. [PubMed]

24. Tabushi, I. Cyclodextrin Catalysis as a Model for Enzyme Action. *Acc. Chem. Res.* **1982**, *15*, 66–72. [CrossRef]

25. Macaev, F.; Boldescu, V. Cyclodextrins in asymmetric and stereospecific synthesis. *Symmetry* **2015**, *7*, 1699–1720. [CrossRef]

26. D'Souza, V.T. Modification of cyclodextrins for use as artificial enzymes. *Supramol. Chem.* **2003**, *15*, 221–229. [CrossRef]

27. Bicchi, C.; Balbo, C.; D'Amato, A.; Manzin, V.; Schreier, P.; Rozenblum, A.; Brunerie, P. Cyclodextrin derivatives in GC separation of racemic mixtures of volatiles - Part XIV: Some applications of thick-film wide-bore columns to enantiomer GC micropreparation. *Hrc-J. High Resolut. Chromatogr.* **1998**, *21*, 103–106. [CrossRef]

28. Armstrong, D.W.; Ward, T.J.; Armstrong, R.D.; Beesley, T.E. Separation of drug stereoisomers by the formation of β-cyclodextrin inclusion complexes. *Science* **1986**, *232*, 1132–1135. [CrossRef] [PubMed]

29. Przybyłek, M.; Cysewski, P. Distinguishing Cocrystals from Simple Eutectic Mixtures: Phenolic Acids as Potential Pharmaceutical Coformers. *Cryst. Growth Des.* **2018**, *18*, 3524–3534. [CrossRef]

30. Steffen, A.; Karasz, M.; Thiele, C.; Lengauer, T.; Kämper, A.; Wenz, G.; Apostolakis, J. Combined similarity and QSPR virtual screening for guest molecules of β-cyclodextrin. *New J. Chem.* **2007**, *31*, 1941–1949. [CrossRef]

31. Linden, L.; Goss, K.U.; Endo, S. 3D-QSAR predictions for α-cyclodextrin binding constants using quantum mechanically based descriptors. *Chemosphere* **2017**, *169*, 693–699. [CrossRef]

32. Katritzky, A.R.; Fara, D.C.; Yang, H.; Karelson, M.; Suzuki, T.; Solov'ev, V.P.; Varnek, A. Quantitative Structure–Property Relationship Modeling of β-Cyclodextrin Complexation Free Energies. *J. Chem. Inf. Comput. Sci.* **2004**, *44*, 529–541. [CrossRef] [PubMed]

33. Prakasvudhisarn, C.; Wolschann, P.; Lawtrakul, L. Predicting complexation thermodynamic parameters of β-cyclodextrin with chiral guests by using swarm intelligence and support vector machines. *Int. J. Mol. Sci.* **2009**, *10*, 2107–2121. [CrossRef] [PubMed]

34. Pérez-Garrido, A.; Helguera, A.M.; Cordeiro, M.N.D.S.; Escudero, A.G. QSPR modelling with the topological substructural molecular design approach: β-cyclodextrin complexation. *J. Pharm. Sci.* **2009**, *98*, 4557–4576. [CrossRef] [PubMed]

35. Rama Krishna, G.; Ukrainczyk, M.; Zeglinski, J.; Rasmuson, Å.C. Prediction of Solid State Properties of Cocrystals Using Artificial Neural Network Modeling. *Cryst. Growth Des.* **2018**, *18*, 133–144. [CrossRef]

36. Zhokhova, N.I.; Bobkov, E.V.; Baskin, I.I.; Palyulin, V.A.; Zefirov, A.N.; Zefirov, N.S. Calculation of the stability of β-cyclodextrin complexes of organic compounds using the QSPR approach. *Moscow Univ. Chem. Bull.* **2007**, *62*, 269–272. [CrossRef]

37. Blanford, W.J.; Gao, H.; Dutta, M.; Ledesma, E.B. Solubility enhancement and QSPR correlations for polycyclic aromatic hydrocarbons complexation with α, β, and γ cyclodextrins. *J. Incl. Phenom. Macrocycl. Chem.* **2014**, *78*, 415–427. [CrossRef]

38. Mirrahimi, F.; Salahinejad, M.; Ghasemi, J.B. QSPR approaches to elucidate the stability constants between β-cyclodextrin and some organic compounds: Docking based 3D conformer. *J. Mol. Liq.* **2016**, *219*, 1036–1043. [CrossRef]

39. Veselinović, A.M.; Veselinović, J.B.; Toropov, A.A.; Toropova, A.P.; Nikolić, G.M. In silico prediction of the β-cyclodextrin complexation based on Monte Carlo method. *Int. J. Pharm.* **2015**, *495*, 404–409. [CrossRef]

40. Friedman, J.H. Multivariate Adaptive Regression Splines. *Ann. Stat.* **1991**, *19*, 1–67. [CrossRef]

41. Przybyłek, M.; Recki, Ł.; Mroczyńska, K.; Jeliński, T.; Cysewski, P. Experimental and theoretical solubility advantage screening of bi-component solid curcumin formulations. *J. Drug Deliv. Sci. Technol.* **2019**, *50*, 125–135. [CrossRef]

42. Przybyłek, M.; Jeliński, T.; Cysewski, P. Application of Multivariate Adaptive Regression Splines (MARSplines) for Predicting Hansen Solubility Parameters Based on 1D and 2D Molecular Descriptors Computed from SMILES String. *J. Chem.* **2019**, *2019*, 1–15. [CrossRef]

43. Suzuki, T. A Nonlinear Group Contribution Method for Predicting the Free Energies of Inclusion Complexation of Organic Molecules with α- and β-Cyclodextrins. *J. Chem. Inf. Comput. Sci.* **2001**, *41*, 1266–1273. [CrossRef] [PubMed]

44. Dong, J.; Cao, D.S.; Miao, H.Y.; Liu, S.; Deng, B.C.; Yun, Y.H.; Wang, N.N.; Lu, A.P.; Zeng, W.B.; Chen, A.F. ChemDes: An integrated web-based platform for molecular descriptor and fingerprint computation. *J. Cheminform.* **2015**, *7*. [CrossRef] [PubMed]

45. Dong, J.; Yao, Z.J.; Wen, M.; Zhu, M.F.; Wang, N.N.; Miao, H.Y.; Lu, A.P.; Zeng, W.B.; Cao, D.S. BioTriangle: A web-accessible platform for generating various molecular representations for chemicals, proteins, DNAs/RNAs and their interactions. *J. Cheminform.* **2016**, *8*. [CrossRef] [PubMed]

46. ChemDes. Available online: http://www.scbdd.com/chemdes/ (accessed on 1 June 2019).

47. BioTriangle. Available online: http://biotriangle.scbdd.com (accessed on 1 June 2019).

48. Ballabio, D.; Consonni, V.; Mauri, A.; Claeys-Bruno, M.; Sergent, M.; Todeschini, R. A novel variable reduction method adapted from space-filling designs. *Chemom. Intell. Lab. Syst.* **2014**, *136*, 147–154. [CrossRef]

49. Ambure, P.; Aher, R.B.; Gajewicz, A.; Puzyn, T.; Roy, K. "NanoBRIDGES" software: Open access tools to perform QSAR and nano-QSAR modeling. *Chemom. Intell. Lab. Syst.* **2015**, *147*, 1–13. [CrossRef]

50. Kennard, R.W.; Stone, L.A. Computer Aided Design of Experiments. *Technometrics* **1969**, *11*, 137–148. [CrossRef]

51. Martin, T.M.; Harten, P.; Young, D.M.; Muratov, E.N.; Golbraikh, A.; Zhu, H.; Tropsha, A. Does rational selection of training and test sets improve the outcome of QSAR modeling? *J. Chem. Inf. Model.* **2012**, *52*, 2570–2578. [CrossRef]

52. QSAR Model Development Using DTC Lab. Software Tools. Available online: http://teqip.jdvu.ac.in/QSAR_Tools/ (accessed on 1 June 2019).

53. Statsoft. *Statistica*; Version 12; StatSoft: Tulsa, OK, USA, 2012.

54. Gramatica, P.; Cassani, S.; Chirico, N. QSARINS-chem: Insubria datasets and new QSAR/QSPR models for environmental pollutants in QSARINS. *J. Comput. Chem.* **2014**, *35*, 1036–1044. [CrossRef]

55. Gramatica, P.; Chirico, N.; Papa, E.; Cassani, S.; Kovarich, S. QSARINS: A new software for the development, analysis, and validation of QSAR MLR models. *J. Comput. Chem.* **2013**, *34*, 2121–2132. [CrossRef]

56. QSAR Research Unit in Environmental Chemistry and Ecotoxicology. Available online: http://www.qsar.it/ (accessed on 1 June 2019).

57. Przybyłek, M.; Jeliński, T.; Słabuszewska, J.; Ziółkowska, D.; Mroczyńska, K.; Cysewski, P. Application of Multivariate Adaptive Regression Splines (MARSplines) Methodology for Screening of Dicarboxylic Acid Cocrystal Using 1D and 2D Molecular Descriptors. *Cryst. Growth Des.* **2019**. [CrossRef]

58. Todeschini, R. Data correlation, number of significant principal components and shape of molecules. The K correlation index. *Anal. Chim. Acta* **1997**, *348*, 419–430. [CrossRef]

59. Todeschini, R.; Consonni, V.; Maiocchi, A. The K correlation index: Theory development and its application in chemometrics. *Chemom. Intell. Lab. Syst.* **1999**, *46*, 13–29. [CrossRef]

60. Stanton, D.T.; Jurs, P.C. Development and Use of Charged Partial Surface Area Structural Descriptors in Computer-Assisted Quantitative Structure-Property Relationship Studies. *Anal. Chem.* **1990**, *62*, 2323–2329. [CrossRef]

61. Labute, P. A widely applicable set of descriptors. *J. Mol. Graph. Model.* **2000**, *18*, 464–477. [CrossRef]

62. Hall, L.H.; Mohney, B.; Kier, L.B. The Electrotopological State: Structure Information at the Atomic Level for Molecular Graphs. *J. Chem. Inf. Comput. Sci.* **1991**, *31*, 76–82. [CrossRef]

63. Hall, L.H.; Kier, L.B. The Molecular Connectivity Chi Indexes and Kappa Shape Indexes in Structure-Property Modeling. *Rev. Comput. Chem.* **1991**, *2*, 367–422.

64. Noolvi, M.N.; Patel, H.M. A comparative QSAR analysis and molecular docking studies of quinazoline derivatives as tyrosine kinase (EGFR) inhibitors: A rational approach to anticancer drug design. *J. Saudi Chem. Soc.* **2013**, *17*, 361–379. [CrossRef]

65. Ji, H.F.; Kong, D.X.; Shen, L.; Chen, L.L.; Ma, B.G.; Zhang, H.Y. Distribution patterns of small-molecule ligands in the protein universe and implications for origin of life and drug discovery. *Genome Biol.* **2007**, *8*. [CrossRef]

66. Bhatiya, R.; Vaidya, A.; Kashaw, S.K.; Jain, A.K.; Agrawal, R.K. QSAR analysis of furanone derivatives as potential COX-2 inhibitors: kNN MFA approach. *J. Saudi Chem. Soc.* **2014**, *18*, 977–984. [CrossRef]

67. Veerasamy, R.; Subramaniam, D.K.; Chean, O.C.; Ying, N.M. Designing hypothesis of substituted benzoxazinones as HIV-1 reverse transcriptase inhibitors: QSAR approach. *J. Enzyme Inhib. Med. Chem.* **2012**, *27*, 693–707. [CrossRef]

68. Ajmani, S.; Janardhan, S.; Viswanadhan, V.N. Toward a general predictive QSAR model for gamma-secretase inhibitors. *Mol. Divers.* **2013**, *17*, 421–434. [CrossRef]

69. Todeschini, R.; Consonni, V. *Molecular Descriptors for Chemoinformatics: Volume 1&2*; Wiley-VCH Verlag GmbH & Co. KGaA: Weinheim, Germany, 2009; ISBN 978-3-527-31852-0.

70. Wen, X.; Tan, F.; Jing, Z.; Liu, Z. Preparation and study the 1:2 inclusion complex of carvedilol with β-cyclodextrin. *J. Pharm. Biomed. Anal.* **2004**, *34*, 517–523. [CrossRef]

71. Kano, K.; Nishiyabu, R.; Asada, T.; Kuroda, Y. Static and dynamic behavior of 2:1 inclusion complexes of cyclodextrins and charged porphyrins in aqueous organic media. *J. Am. Chem. Soc.* **2002**, *124*, 9937–9944. [CrossRef] [PubMed]

72. Frixa, C.; Scobie, M.; Black, S.J.; Thompson, A.S.; Threadgill, M.D. Formation of a remarkably robust 2:1 complex between β-cyclodextrin and a phenyl-substituted icosahedral carborane. *Chem. Commun.* **2002**, *2*, 2876–2877. [CrossRef]

73. U.S. Food and Drug Administration The Biopharmaceutics Classification System (BCS) Guidance. Available online: https://www.fda.gov/ (accessed on 1 June 2019).

74. Loftsson, T. Cyclodextrins and the biopharmaceutics classification system of drugs. *J. Incl. Phenom.* **2002**, *44*, 63–67. [CrossRef]

75. Loftsson, T. Drug permeation through biomembranes: Cyclodextrins and the unstirred water layer. *Pharmazie* **2012**, *67*, 363–370.

76. Dahan, A.; Wolk, O.; Kim, Y.H.; Ramachandran, C.; Crippen, G.M.; Takagi, T.; Bermejo, M.; Amidon, G.L. Purely in silico BCS classification: Science based quality standards for the world's drugs. *Mol. Pharm.* **2013**, *10*, 4378–4390. [CrossRef]

77. Xiao, D.; Pham-Huy, L.A.; Pham-Huy, C.; Dramou, P.; He, H.; Zuo, P. Carbon Nanotubes: Applications in Pharmacy and Medicine. *Biomed Res. Int.* **2013**, *2013*, 1–12.

78. Singh, I.; Rehni, A.K.; Kumar, P.; Kumar, M.; Aboul-Enein, H.Y. Carbon Nanotubes: Synthesis, Properties and Pharmaceutical Applications. *Fullerenes Nanotub. Carbon Nanostruct.* **2009**, *17*, 361–377. [CrossRef]

79. Szefler, B. Nanotechnology, from quantum mechanical calculations up to drug delivery. *Int. J. Nanomed.* **2018**, *13*, 6143–6176. [CrossRef]

Article

DFT Calculations of the Structural, Mechanical, and Electronic Properties of TiV Alloy Under High Pressure

Fang Yu [1] and Yu Liu [2,*]

[1] School of Software and Communication Engineering, Xiangnan University, Chenzhou 423000, China
[2] State Key Laboratory of Advanced Design and Manufacturing for Vehicle Body, Hunan University, Changsha 410082, China
* Correspondence: lyu1006@hnu.edu.cn

Received: 8 July 2019; Accepted: 29 July 2019; Published: 1 August 2019

Abstract: A calculation program based on the density functional theory (DFT) is applied to study the structural, mechanical, and electronic properties of TiV alloys with symmetric structure under high pressure. We calculate the dimensionless ratio, elastic constants, shear modulus, Young's modulus, bulk modulus, ductile-brittle transition, material anisotropy, and Poisson's ratio as functions of applied pressure. Results suggest that the critical pressure of structural phase transition is 42.05 GPa for the TiV alloy, and structural phase transition occurs when the applied pressure exceeds 42.05 GPa. High pressure can improve resistance to volume change, as well as the ductility and atomic bonding, but the strongest resistances to elastic and shear deformation occur at $P = 5$ GPa for TiV alloy. Furthermore, the results of the density of states (DOS) indicate that the TiV alloy presents metallicity. High pressure disrupts the structural stability of the TiV alloy with symmetry, thereby inducing structural phase transition.

Keywords: TiV alloy; Symmetric structure; DFT calculation; mechanical property; elastic constant; electronic structure; applied pressure

---

1. Introduction

TiV alloys with high gravimetric and volumetric hydrogen storage capacities have been widely regarded as important hydrogen storage materials [1–4]. Iba and Akiba [5] reported TiV–Mn alloys with multiphase nanostructures of body-centered cubic (BCC) and C14-type Laves phases, which had a large hydrogen capacity and excellent desorbing properties. Then, they revealed that the improved hydrogen sorption properties were attributed to the generation of multiphase nanostructures. Yu et al. [6] studied the hydrogen storage performance of a single BCC phase Ti-40V-10Cr-10Mn alloy, and results indicated that the largest hydrogen absorption capacity of the Ti–V–Cr–Mn alloy can reach 4.2 wt%, exceeding that of other hydrogen storage alloys, such as rare-earth (RE)-, Ti-, and Zr-based alloys. Nomura and Akiba [7] also investigated 26 types of alloys composed of Ti(33–47 mol%)-V(42–67 mol%)-Fe(0–14 mol%), and then found that the most suitable proportion for hydrogen absorption was $Ti_{43.5}V_{49.0}Fe_{7.5}$, and the hydrogen capacity reached up to 3.90 wt% (H/M = 1.90) at 253 K. Meanwhile, Seo et al. [8] reported that the $V_{0.68}Ti_{0.20}Fe_{0.12}$ alloy showed the largest hydrogen capacity of 3.6 wt%, and heat treatment can effectively improve hydrogen capacity, such as that in the $V_{0.375}Ti_{0.20}Cr_{0.30}Mn_{0.075}$ alloy with a hydrogen capacity of 2.2 wt%. Therefore, as the new hydrogen storage materials, TiV-based alloys have been greatly investigated [4,9–11]. However, the structural, mechanical, and electronic properties of TiV alloys remain unrevealed under high pressure due to the complexity of calculations, thereby limiting the applications of TiV alloys under high pressure.

Thus, this work uses first-principle calculations to study the structural, mechanical, and electronic properties of the TiV alloy with symmetry systematically within the frame of the density functional theory (DFT) and computes in detail the some variables with respect to the different pressures, such as the dimensionless ratio, elastic constants, elastic modulus, ductile–brittle transition, anisotropy factors, and Poisson's ratio, along with electronic properties. Moreover, the calculated results agree well with other experimental data and theoretical results. Therefore, the results can provide valuable guidance for the development and application of TiV alloys in the area of hydrogen storage materials.

2. Methodology

In the present work, all DFT calculations were performed by the Cambridge Serial Total Energy Package Program [12–14], which aims to obtain the electronic and energy properties of each structure in TiV alloys. The exchange–correlation function was decided by the generalized gradient approximation of Perdew–Burke–Ernzerhof [15]. The electronic states of Ti ($4s^23p^63d^2$) and V ($4s^23p^63d^3$) are the valence electrons, and the Vanderbilt-type ultrasoft pseudopotentials were applied to address the ion–electron interactions [16]. On the basis of a precise convergence test, the plane–wave cutoff energy was optimized as 400 eV, and the Brillouin-zone k-point grid [17] was selected as $13 \times 13 \times 13$ in the electronic calculations of the TiV alloy. The space group of the TiV alloy belongs to *Im-3m*, and the symmetric crystal structure of the TiV alloy is shown in Figure 1, and the Broyden–Fletcher–Goldfarb–Shanno (BFGS) algorithm [18] was applied in the process of structural geometry optimization with respect to the applied pressures ranging from –10 to 50 GPa. The energy convergence criterion was set at 1.0×10^{-6} eV/atom in self-consistent calculations, and the Hellmann–Feynman force of each atom was lower than 0.01 eV/Å.

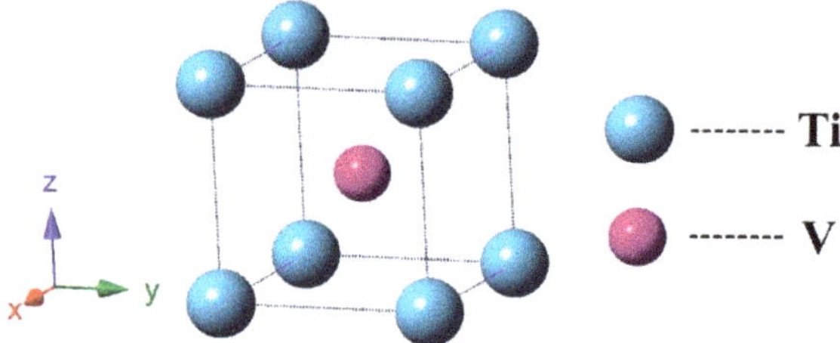

Figure 1. Symmetric Crystal structure of the TiV alloy.

3. Results and Discussion

3.1. Structural Properties and Stability

Prior to studying the various properties of the TiV alloy, the most stable crystal structure of the TiV alloy was first obtained, and the $E - V$ data of total energy E with respect to volume V were calculated, and then they were accurately fitted by the Birch–Murnaghan equation of state [19], as shown in Figure 2, where the variation ranges of volume V are from 0.9 V_0 to 1.1 V_0. During the geometry optimization of TiV alloys, the full optimization for a unit cell was implemented to obtain the total energy of the stable state. Figure 2 shows that for volume $V = V_0$ ($V_0 = 29.971$ Å^3), the total energy reached the minimum value ($E_t = -16.457$ eV), and the lattice constant was accordingly $a_0 = 3.107$ Å, indicating the structural parameters of the TiV alloy under the most stable structure, where V_0 is the primitive cell volume, and a_0 is the equilibrium lattice constant at pressure $P = 0$ and temperature $T = 0$, respectively. The present results agree well with other approaches [10,20–25] and are comparatively listed in Table 1.

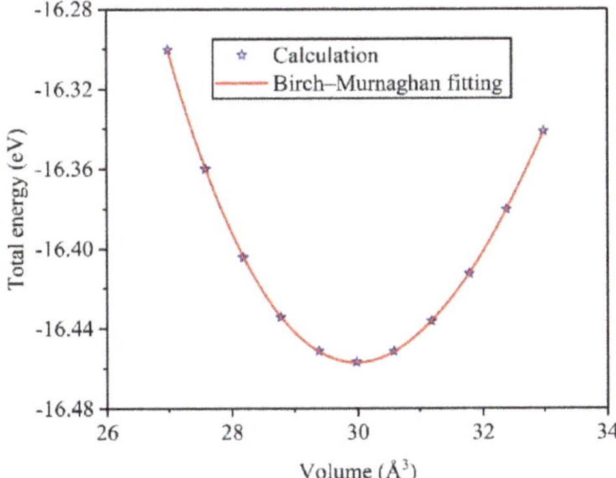

Figure 2. The $E - V$ curve of total energy E with respect to cell volume V of the TiV alloy.

Table 1. Comparisons of the calculated lattice constant to the ones of others for TiV alloy.

TiV Alloy	This Work	Others
Lattice constant a (Å)	3.107	3.156 [10], 3.163 [20], 3.165 [21] 3.120 [22], 3.159 [23], 3.140 [24] 3.280 [25]

To investigate the dependencies of the lattice constant and unit cell volume on the applied pressure, a series of structural optimizations were conducted to obtain the corresponding lattice constant under various applied pressures. Then, the dimensionless ratios a/a_0 and V/V_0 with respect to the different applied pressures, which are depicted in Figure 3, decreased monotonously with an increment of applied pressure, and the compression ratio of the volume was much larger compared with the lattice constant at the same applied pressure, thereby indicating that high pressure greatly reduces interatomic distance and leads to strong electron interactions.

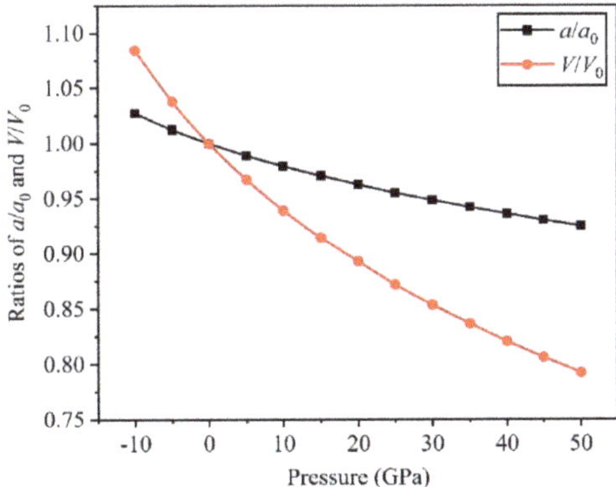

Figure 3. Dependencies of dimensionless ratios a/a_0 and V/V_0 on the applied pressure for the TiV alloy.

In Figure 3, we fit the dependencies of dimensionless ratios a/a_0 and V/V_0 as a function of the applied pressure by using quadratic polynomial, and they are expressed as follows:

$$a/a_0 = 1.001 - 2.240 \times 10^{-3}P + 1.501 \times 10^{-5}P^2 \tag{1}$$

$$V/V_0 = 1.005 - 6.690 \times 10^{-3}P + 5.087 \times 10^{-5}P^2 \tag{2}$$

In the anisotropic material, the elastic constants, which denote the ability to resist applied stress, are considered important physical quantities in measuring the structural stability of a material. For cubic crystals, the elastic constants include three coefficients, namely, C_{11}, C_{12}, and C_{44}. In accordance with the stability criterion [26,27], three elastic constants of cubic crystals fit with the following expressions:

$$(C_{11} - C_{12}) > 0, \; C_{11} > 0, \; C_{44} > 0, \; (C_{11} + 2C_{12}) > 0 \tag{3}$$

At $P = 0$ and $T = 0$, Table 2 lists the calculated elastic constants, bulk modulus B, Young's modulus E, and shear modulus G, as well as Poisson's ratio σ. Table 2 shows that the calculated results are consistent with the work of Ikehata et al. [25].

Table 2. Comparisons of the calculated results with the work of Ikehata et al. [25] at $P = 0$ and $T = 0$.

TiV Alloy	This Work	Ikehata et al. [25]
C_{11} (GPa)	178.16	169.6
C_{12} (GPa)	126.92	122.3
C_{44} (GPa)	21.52	33.6
C_{44} (GPa)	144.00	138.07
Young's modulus E (GPa)	65.71	81.81
Shear modulus G (GPa)	23.07	29.19
Poisson's ratio σ	0.42	0.40

Figure 4 demonstrates the dependencies of elastic constants C_{11}, C_{12}, and C_{44} on the applied pressure for the TiV alloy, revealing that for the pressure P changing from 0 GPa to 42.05 GPa, C_{11} and C_{12} increase gradually, but the elastic coefficient C_{44} decreases. Furthermore, the elastic coefficient C_{44} reduces slowly to zero with increasing applied pressure and then dramatically transforms into a negative value under the high pressure exceeding 42.05 GPa, which is unfit for the stability criterion of Equation (3). Thus, a high pressure exceeding 42.05 GPa disrupts the crystalline structure of the TiV alloy, namely, the TiV alloy is inclined to induce structural phase transition in the case of a high pressure of 42.05 GPa.

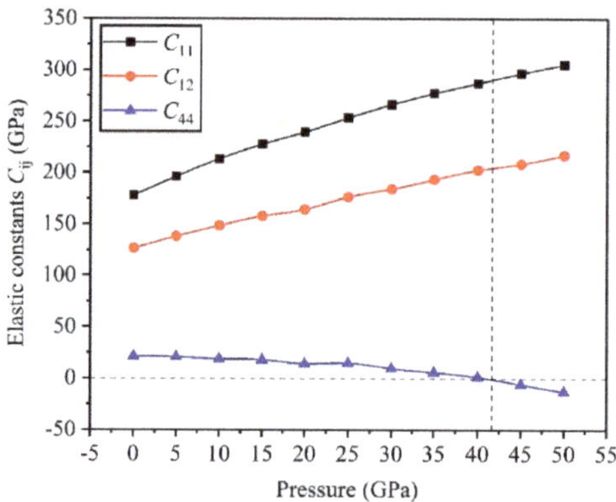

Figure 4. Dependencies of the elastic constants C_{ij} on the applied pressure for the TiV alloy.

3.2. Mechanical Properties

The mechanical properties of materials, such as high strength, good plasticity, and excellent deformation resistance, are closely related to material moduli, like shear modulus G, Young's modulus E, and bulk modulus B [28,29]. A large modulus has great resistance to material deformation.

Furthermore, we can apply Equations (4)–(6) [30] to calculate the modulus values of these materials. These modulus values depend on the calculated elastic constants, as listed in Table 2:

$$B = \frac{1}{3}(C_{11} + 2C_{12}) \tag{4}$$

$$G = \frac{1}{2}(G_V + G_R) \tag{5}$$

$$E = \frac{9BG}{3B + G} \tag{6}$$

where $G_V = (C_{11} - C_{12} + 3C_{44})/5$ and $G_R = 5(C_{11} - C_{12})C_{44}/[4C_{44} + 3(C_{11} - C_{12})]$, and G_V and G_R denote the Voigt and Reuss shear moduli, respectively.

Figure 5 demonstrates the dependencies of shear modulus G, Young's modulus E, and bulk modulus B on the applied pressure. Prior to structural phase transition, the bulk modulus B increased with increasing applied pressure, indicating that the high pressure leads to a strong resistance to volume change for the TiV alloy, and the maximum bulk modulus B_{max} is equal to 233.90 GPa at $P = 42.05$ GPa. However, shear modulus G and Young's modulus E initially increased from 0 GPa to 5 GPa and then decrease slowly with increasing applied pressure. The maximum values of Young's modulus E and shear modulus G were 68.47 and 23.98 GPa at $P = 5$ GPa, respectively, implying that the applied pressure at this value generates the maximum resistance to elastic and shear deformations for the TiV alloy.

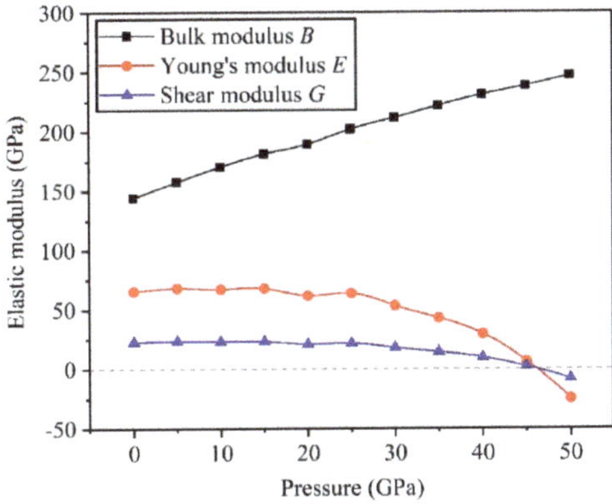

Figure 5. Dependencies of shear modulus G, Young's modulus E, and bulk modulus B on the applied pressure for the TiV alloy.

To study the structural phase transition of materials, the present work aimed to analyze the ductile–brittle transition under different pressures. Pugh [28] raised that the modulus ratio B/G can be considered the key physical quantities in measuring the ductile/brittle properties for polycrystalline materials. Generally, 1.75 is the critical value of the ratio B/G for distinguishing ductile and brittle materials. The materials are characterized for the ductility by $B/G > 1.75$, and they are characterized for the brittleness by $B/G < 1.75$. Moreover, the ductility of materials increases with increasing B/G ratio. Otherwise, the brittleness increases. These conditions are equally suitable for investigating the ductile–brittle transition of intermetallic compounds [31]. Figure 6 exhibits the calculated results of the present work, showing that the modulus ratio B/G is approximately 6.24 under $P = 0$ GPa, and the B/G ratio (> 1.75) increases with an increase of applied pressure. Thus, the results from Figure 6 indicate that the TiV alloy is essentially characterized by excellent ductility, and high pressure can improve the ductility of the TiV alloy.

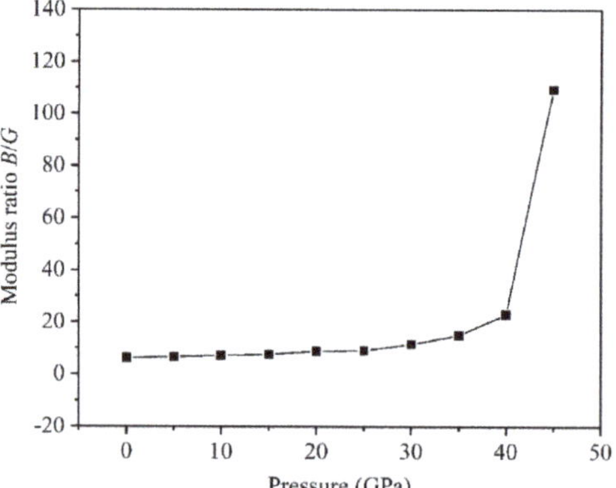

Figure 6. Dependencies of the modulus ratio B/G on the applied pressure for the TiV alloy.

3.3. Anisotropy

For anisotropic materials, we used elastic anisotropy to study their mechanical properties. The anisotropy factor A plays a key role in measuring the elastic anisotropy of materials, which are characterized by the isotropy at $A = 1$. Otherwise, they are characterized by the anisotropy at $A \neq 1$. Various anisotropy factors A ($A \neq 1$) correspond to the varying degrees of material anisotropy, and the material anisotropy is stronger when the anisotropy factors A are more deviated [32,33]. In the light of the work of Yoo [34], the cross-slip pinning model was presented to investigate the cross-slip pinning process of screw dislocations. The large factor A can increase the driving force of screw dislocation motion, and then promote the cross-slip pinning process of such dislocations. Here, we calculated the anisotropy factors $A_{(100)[001]}$ and $A_{(110)[001]}$ using Equations (7) and (8), respectively. They are expressed by the three elastic constants of cubic crystals, as follows [35,36]:

$$A_{(100)[001]} = \frac{2C_{44}}{C_{11} - C_{12}} \tag{7}$$

$$A_{(110)[001]} = \frac{C_{44}(C' + 2C_{12} + C_{11})}{C_{11}C' - C_{12}{}^2} \tag{8}$$

where $C' = C_{44} + (C_{11} + C_{12})/2$, $A_{(100)[001]}$ is the anisotropy factor in the (100)[001] direction, and $A_{(110)[001]}$ in the (110)[001] directions. Based on the calculations of Equations (7) and (8), Figure 7 shows the dependencies of the two anisotropy factors on the applied pressure. These factors helped investigate the mechanical properties of the TiV alloy under various pressures. Results in Figure 7 indicate that the anisotropy factors of the TiV alloy are $A_{(100)[001]} = 0.840$ and $A_{(110)[001]} = 0.875$ at $P = 0$ GPa, suggesting that the TiV alloy is essentially characterized by the anisotropy. As the applied pressure increases, the anisotropy factors $A_{(100)[001]}$ and $A_{(110)[001]}$ decrease rapidly, indicating that a larger applied pressure leads to a stronger anisotropy for the TiV alloy. Moreover, the cross-slip process of screw dislocations can be accelerated due to the strong anisotropy caused by the high pressure.

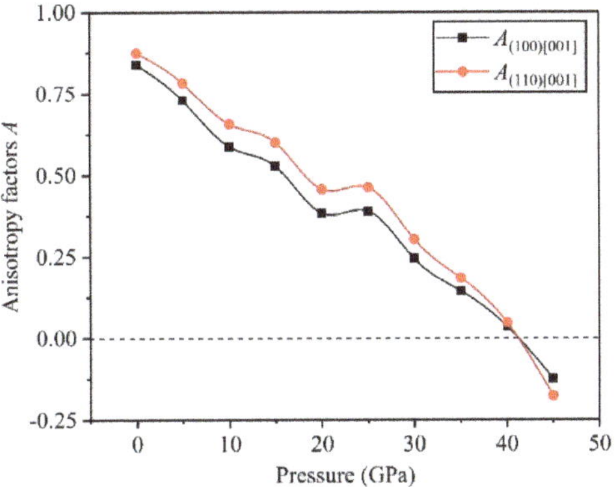

Figure 7. Dependencies of anisotropy factors on the applied pressure for the TiV alloy.

To investigate the plastic property of the TiV alloy under different pressures, we introduced the Poisson's ratio σ. Generally, the range of Poisson's ratio σ is between -1 and 0.5, and the material has better plasticity when Poisson's ratio σ is larger. Based on the key factors of atomic scale, Poisson's ratio σ mainly depended on the type of interatomic bonding [36]. In the work of Reed and Clark [37], Poisson's ratios $\sigma_{min} = 0.25$ and $\sigma_{max} = 0.5$ were identified as the minimum and maximum values in investigating central force solids, respectively. Herein, we define the $\sigma_{[001]}$ and $\sigma_{[111]}$ as the Poisson's ratios in the directions of [001] and [111], respectively, and they were obtained using Equations (9) and (10) [36,38]. Figure 8 depicts the dependencies of Poisson's ratios $\sigma_{[001]}$ and $\sigma_{[111]}$ on the applied pressure. We can obtain the values of $\sigma_{[001]} = 0.416$ and $\sigma_{[111]} = 0.429$ at $P = 0$ GPa. Thus, for the TiV alloy in nature, interatomic bonding is mainly characterized by the central force in the directions of [001] and [111]. As the applied pressure increases, the value of $\sigma_{[001]}$ initially decreases and then increases, but that of $\sigma_{[111]}$ gradually increases and then tends to achieve the maximum value of $\sigma_{max} = 0.5$, thereby indicating that the high pressure enhances the central force and plasticity of the TiV alloy in the [111] uniaxial direction.

$$\sigma_{[001]} = \frac{C_{12}}{C_{11} + C_{12}} \tag{9}$$

$$\sigma_{[111]} = \frac{C_{11} + 2C_{12} - 2C_{44}}{2(C_{11} + 2C_{12} + C_{44})} \tag{10}$$

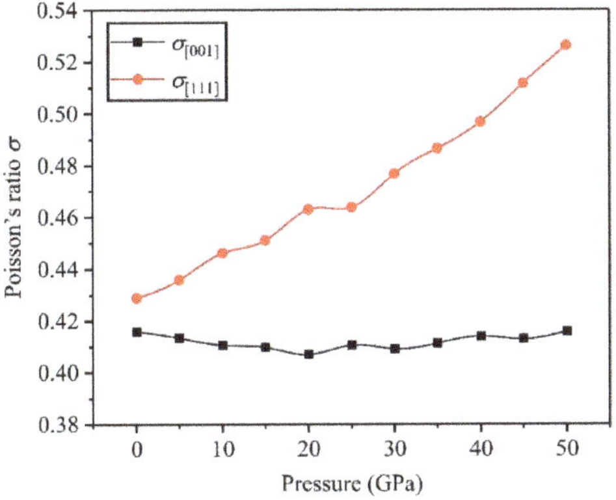

Figure 8. Dependencies of Poisson's ratios on the applied pressure for the TiV alloy.

The mechanical properties of the materials can also be evaluated by other physical quantities. The evaluation provides valuable guidance for studying material strength and deformation. Variables $G_{(100)[010]}$ and $G_{(110)[1\bar{1}0]}$ are the shear moduli in the $(100)[010]$ and $(110)[1\bar{1}0]$ directions, respectively, where they are expressed by $G_{(100)[010]} = C_{44}$ and $G_{(110)[1\bar{1}0]} = (C_{11} - C_{12})/2$. Similarly, variable $E_{\langle 100 \rangle}$ is the Young's modulus in the $\langle 100 \rangle$ uniaxial directions, and $E_{\langle 100 \rangle} = (C_{11} - C_{12})[1 + C_{12}/(C_{11} + C_{12})]$ [33, 39]. According to the definitions of these moduli, Figure 9 exhibits their dependencies on the applied pressure. Results show that shear modulus $G_{(110)[1\bar{1}0]}$ is always greater than $G_{(100)[010]}$ under any pressure, thereby indicating that the $(110)[1\bar{1}0]$ direction has stronger resistance to shear deformation compared with the $(100)[010]$ direction. Moreover, $G_{(110)[1\bar{1}0]}$ increases gradually with increasing applied pressure, thereby revealing that the high pressure enhances the resistance to shear deformation in the $(110)[1\bar{1}0]$ direction. However, the high pressure reduces the resistance to shear deformation in the $(100)[010]$ direction due to the decrement of $G_{(100)[010]}$. Young's modulus $E_{\langle 100 \rangle}$ increases gradually with increasing applied pressure. Thus, high pressure improves the resistance to elastic deformation in the $\langle 100 \rangle$ directions. As another physical quantity, the Cauchy pressure $C_{12} - C_{44}$ is generally used to reflect the characteristic of atomic bonding, and it can reveal the nature of the bonding from the atomic scale [40]. A positive Cauchy pressure denotes that the atomic bonding exhibits metallic properties. Thus, a homogeneous electron gas exists near the spherical atoms, and the electron distribution shows the same characteristics without regionality and directionality. However, a negative Cauchy pressure is characterized by the directional bonding, and the directional characteristic increases with increasing Cauchy pressure [33,41]. Results in Figure 9 show that the value of Cauchy pressures $C_{12} - C_{44}$ is always positive under any pressure and increases linearly with increasing applied pressure, thereby suggesting that the atomic bonding of the TiV alloy is mainly the metallic bond, and high pressure leads to a strong atomic bonding.

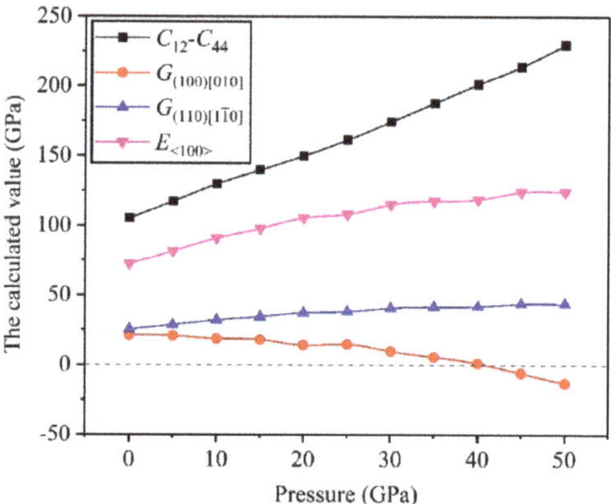

Figure 9. Dependencies of material moduli on the applied pressure for the TiV alloy.

3.4. Electronic Properties

To further investigate the bonding mechanisms of the TiV alloy, we obtained the electronic structures to reveal the stability changes of the TiV alloy under different pressures. Figure 10 depicts the partial density of states (PDOS) and total DOS (TDOS) of the TiV alloy under $P = 0$ GPa, and the Fermi level ($E_F = 0$ eV) is marked by the red dash line. The value of TDOS at E_F was not equal to zero, indicating that the TiV alloy presents the metallicity in nature and agrees well with the results in Figure 9. In light of the PDOS of the TiV alloy, the contributions to DOS are mainly obtained from the Ti-3d and V-3d states at the E_F, and the Ti-4s and V-4s states have no contributions to DOS.

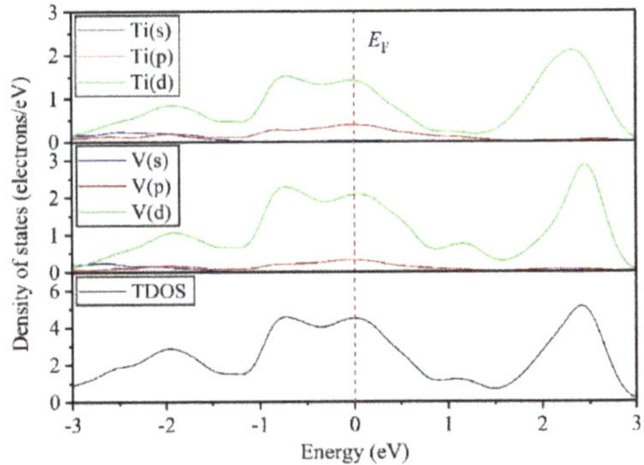

Figure 10. PDOS and TDOS for the TiV alloy under $P = 0\,\mathrm{GPa}$.

Simultaneously, we also investigated the influences of the applied pressure on TDOS of the TiV alloy, and Figure 11 demonstrates the dependencies of the TDOS of the TiV alloy on the applied pressure. With increasing applied pressure, the valence band energy increases but the conduction band energy decreases, thereby resulting in the decrement of the band gap between valence band and conduction band. Then, the stability of the crystal structure decreases to a certain extent. The crystal structure of the TiV alloy easily produces the structural phase transition under high pressure. Results agree well with the previous results.

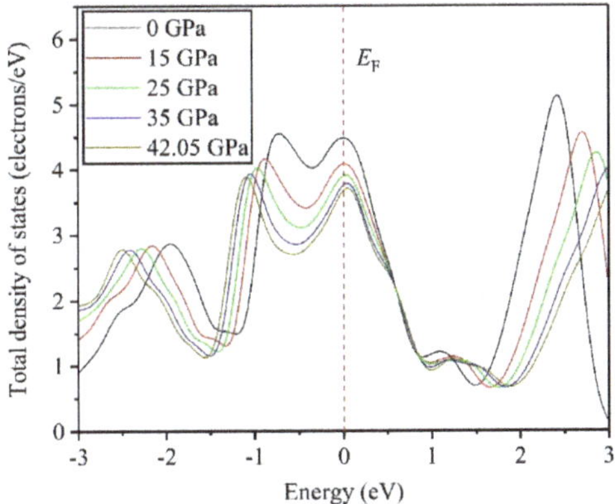

Figure 11. TDOS for the TiV alloy under various pressures.

Figure 12 plots the schematic of the isosurface contours of charge density under different pressures. The figure intuitively reflects the local chemical bonding between Ti and V atoms. Figure 12 shows that the local charge density between Ti and V atoms increases gradually with the increase of applied pressure, suggesting that the electron interaction between Ti and V atoms becomes increasingly strong under high pressure. Meanwhile, the results indirectly reflect that the local chemical bonding combined with the electrons and ions, between Ti and V atoms becomes stronger with increasing the applied pressure as well. The high pressure induces strong interactions between the Ti and V atoms, leading to structural instability under high pressure. Thus, the high pressure tends to destroy the structural stability of the TiV alloy, thereby inducing structural phase transition.

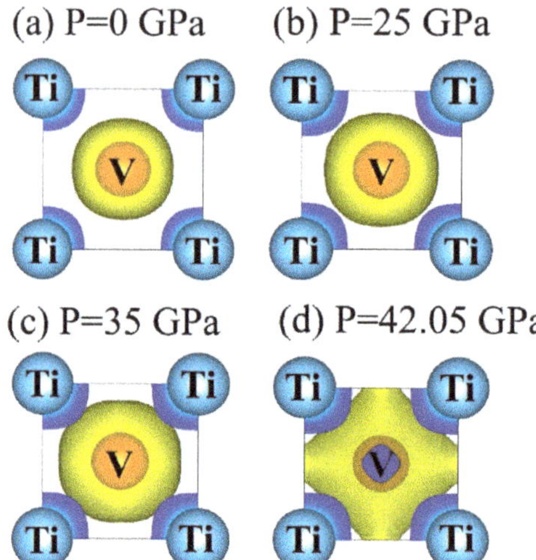

Figure 12. Schematic of isosurface contours of charge density for the TiV alloy under different pressures. The isosurface levels are set as $0.0406\, r_0^{-3}$ (r_0: Bohr radius).

4. Conclusions

(1) The critical pressure of the structural phase transition for the TiV alloy is 42.05 GPa, and the symmetric crystal structure of the TiV alloy produces structural phase transition when the applied pressure exceeds 42.05 GPa.

(2) The high pressure can improve resistance to volume change, but the biggest resistances to elastic and shear deformation occur under $P = 5$ GPa for the TiV alloy.

(3) The results of Pugh's B/G ratio suggest that the TiV alloy is essentially characterized by excellent ductility, and high pressure can enhance the ductility of the TiV alloy.

(4) The $(110)\left[1\bar{1}0\right]$ direction has stronger resistance to shear deformation than the $(100)[010]$ direction, and high pressure improves resistance to elastic deformation in the $\langle 100 \rangle$ direction. Cauchy pressure reveals that the atomic bonding of the TiV alloy is mainly the metallic bond, and high pressure leads to strong atomic bonding.

(5) DOS results indicate that the TiV alloy presents metallicity, and high pressure disrupts the structural stability of the TiV alloy, thereby inducing structural phase transition.

Author Contributions: Conceptualization, F.Y. and Y.L.; methodology, F.Y. and Y.L.; software, F.Y.; validation, F.Y. and Y.L.; formal analysis, F.Y. and Y.L.; investigation, F.Y. and Y.L.; resources, F.Y.; data curation, F.Y. and Y.L.; writing—original draft preparation, F.Y. and Y.L.; writing—review and editing, F.Y. and Y.L.; visualization, F.Y.; supervision, F.Y. and Y.L.; project administration, F.Y. and Y.L.; funding acquisition, F.Y.

Funding: This research was funded by the Cooperative Education Program of Ministry of Education of China (No. 201801238024), Innovation and Entrepreneurship Education Center Project of Hunan Provincial Education Department ([2018]380-74), and Chenzhou Municipal Science and Technology Bureau of Research on Real-Time Monitoring System of Intelligent Trash Can ([2018]102).

Acknowledgments: We deeply appreciate the computing resources offered by the National Supercomputing Center in Shenzhen, China.

Conflicts of Interest: The authors declare no conflict of interest.

References

1. Bibienne, T.; Tousignant, M.; Bobet, J.L.; Huot, J. Synthesis and hydrogen sorption properties of $TiV_{(2-x)}Mn_x$ BCC alloys. *J. Alloy. Compd.* **2015**, *624*, 247–250. [CrossRef]

2. Massicot, B.; Latroche, M.; Joubert, J.M. Hydrogenation properties of Fe–Ti–V bcc alloys. *J. Alloy. Compd.* **2011**, *509*, 372–379. [CrossRef]

3. Huot, J.; Ravnsbæk, D.B.; Zhang, J.; Cuevas, F.; Latroche, M.; Jensen, T.R. Mechanochemical synthesis of hydrogen storage materials. *Prog. Mater. Sci.* **2013**, *58*, 30–75. [CrossRef]

4. Yu, X.B.; Yang, Z.X.; Feng, S.L.; Wu, Z.; Xu, N.X. Influence of Fe addition on hydrogen storage characteristics of Ti–V-based alloy. *Int. J. Hydrog. Energy* **2006**, *31*, 1176–1181. [CrossRef]

5. Iba, H.; Akiba, E. Hydrogen absorption and modulated structure in Ti–V–Mn alloys. *J. Alloy. Compd.* **1997**, *253*, 21–24. [CrossRef]

6. Yu, X.B.; Wu, Z.; Xia, B.J.; Huang, T.Z.; Chen, J.Z.; Wang, Z.S.; Xu, N.X. Hydrogen storage in Ti–V-based body-centered-cubic phase alloys. *J. Mater. Res.* **2003**, *18*, 2533–2536. [CrossRef]

7. Nomura, K.; Akiba, E. H_2 Absorbing-desorbing characterization of the Ti-V-Fe alloy system. *J. Alloy. Compd.* **1995**, *231*, 513–517. [CrossRef]

8. Seo, C.Y.; Kim, J.H.; Lee, P.S.; Lee, J.Y. Hydrogen storage properties of vanadium-based b.c.c. solid solution metal hydrides. *J. Alloy. Compd.* **2003**, *348*, 252–257. [CrossRef]

9. Pan, H.G.; Li, R.; Gao, M.X.; Liu, Y.F.; Wang, Q.D. Effects of Cr on the structural and electrochemical properties of TiV-based two-phase hydrogen storage alloys. *J. Alloy. Compd.* **2005**, *404*, 669–674. [CrossRef]

10. Balcerzak, M. Structure and hydrogen storage properties of mechanically alloyed Ti-V alloys. *Int. J. Hydrog. Energy* **2017**, *42*, 23698–23707. [CrossRef]

11. Dou, T.; Wu, Z.; Mao, J.F.; Xu, N.X. Application of commercial ferrovanadium to reduce cost of Ti–V-based BCC phase hydrogen storage alloys. *Mater. Sci. Eng. A* **2008**, *476*, 34–38. [CrossRef]

12. Milman, V.; Winkler, B.; White, J.A.; Pickard, C.J.; Payne, M.C.; Akhmatskaya, E.V.; Nobes, R.H. Electronic Structure, Properties, and Phase Stability of Inorganic Crystals: A Pseudopotential Plane-Wave Study. *Int. J. Quantum Chem* **2000**, *77*, 895–910. [CrossRef]

13. Segall, M.D.; Lindan, P.J.D.; Probert, M.J.; Pickard, C.J.; Hasnip, P.J.; Clark, S.J.; Payne, M.C. First-Principles Simulation: Ideas, Illustrations and the CASTEP Code. *J. Phys. Condens. Matter* **2002**, *14*, 2717–2744. [CrossRef]

14. Clark, S.J.; Segall, M.D.; Pickard, C.J.; Hasnip, P.J.; Probert, M.I.J.; Refson, K.; Payne, M.C. First principles methods using CASTEP. *Z. Krist. Cryst. Mater.* **2005**, *220*, 567–570. [CrossRef]

15. Perdew, J.P.; Burke, K.; Ernzerhof, M. Generalized Gradient Approximation Made Simple. *Phys. Rev. Lett.* **1996**, *77*, 3865–3868. [CrossRef]

16. Vanderbilt, D. Soft self-consistent pseudopotentials in a generalized eigenvalue formalism. *Phys. Rev. B* **1990**, *41*, 7892–7895. [CrossRef]

17. Monkhorst, H.J.; Pack, J.D. Special points for Brillouin-zone integrations. *Phys. Rev. B* **1976**, *13*, 5188–5192. [CrossRef]

18. Fischer, T.H.; Almlof, J. General methods for geometry and wave function optimization. *J. Phys. Chem.* **1992**, *96*, 9768–9774. [CrossRef]

19. Birch, F. Finite Elastic Strain of Cubic Crystals. *Phys. Rev.* **1947**, *71*, 809–824. [CrossRef]

20. Aurelio, G.; Guillermet, A.F.; Cuello, G.J.; Campo, J. Metastable phases in the Ti-V system: Part I. Neutron diffraction study and assessment of structural properties. *Metall. Mater. Trans. A* **2002**, *33*, 1307–1317. [CrossRef]

21. Ming, L.C.; Manghnani, M.H.; Katahara, K.W. Phase transformations in the Ti-V system under high pressure up to 25 GPa. *Acta Metall.* **1981**, *29*, 479–485. [CrossRef]

22. Basak, S.; Shashikala, K.; Sengupta, P.; Kulshreshtha, S.K. Hydrogen absorption properties of Ti–V–Fe alloys: Effect of Cr substitution. *Int. J. Hydrog. Energy* **2007**, *32*, 4973–4977. [CrossRef]

23. Hagi, T.; Sato, Y.; Yasuda, M.; Tanaka, K. Structure and phase diagram of the Ti-V-H system at room temperature. *Trans. Jpn. Inst. Met.* **1987**, *28*, 198–204. [CrossRef]

24. Stern, A.; Kaplan, N.; Shaltiel, D. Superconducting transition temperatures of the system $V_{1-x}Ti_xH_y$. *Solid State Commun.* **1981**, *38*, 445–450. [CrossRef]

25. Ikehata, H.; Nagasako, N.; Furuta, T.; Fukumoto, A.; Miwa, K.; Saito, T. First-principles calculations for development of low elastic modulus Ti alloys. *Phys. Rev. B* **2004**, *70*, 174113. [CrossRef]

26. Wang, J.H.; Yip, S.; Phillpot, S.R.; Wolf, D. Crystal instabilities at finite strain. *Phys. Rev. Lett.* **1993**, *71*, 4182–4185. [CrossRef]

27. Patil, S.K.R.; Khare, S.V.; Tuttle, B.R.; Bording, J.K.; Kodambaka, S. Mechanical stability of possible structures of PtN investigated using first-principles calculations. *Phys. Rev. B* **2006**, *73*, 104118. [CrossRef]

28. Pugh, S.F. XCII. Relations between the elastic moduli and the plastic properties of polycrystalline pure metals. *Philos. Mag.* **1954**, *45*, 823–843. [CrossRef]

29. Cao, Y.; Zhu, J.C.; Nong, Z.S.; Yang, X.W.; Liu, Y.; Lai, Z.H. First-principles studies of the structural, elastic, electronic and thermal properties of Ni_3Nb. *Comput. Mater. Sci.* **2013**, *77*, 208–213. [CrossRef]

30. Iotova, D.; Kioussis, N.; Lim, S.P. Electronic structure and elastic properties of the Ni_3X (X=Mn, Al, Ga, Si, Ge) intermetallics. *Phys. Rev. B* **1996**, *54*, 14413–14422. [CrossRef]

31. Hill, R. The elastic behaviour of a crystalline aggregate. *Proc. Phys. Soc. Sect A* **1952**, *65*, 349–354. [CrossRef]

32. Mattesini, M.; Ahuja, R.; Johansson, B. Cubic Hf_3N_4 and Zr_3N_4: A class of hard materials. *Phys. Rev. B* **2003**, *68*, 184108. [CrossRef]

33. Fu, H.Z.; Li, D.H.; Peng, F.; Gao, T.; Cheng, X.L. Ab initio calculations of elastic constants and thermodynamic properties of NiAl under high pressures. *Comput. Mater. Sci.* **2008**, *44*, 774–778. [CrossRef]

34. Yoo, M.H. On the theory of anomalous yield behavior of Ni_3Al—Effect of elastic anisotropy. *Scr. Metall.* **1986**, *20*, 915–920. [CrossRef]

35. Lau, K.; Mccurdy, A.K. Elastic anisotropy factors for orthorhombic, tetragonal, and hexagonal crystals. *Phys. Rev. B* **1998**, *58*, 8980–8984. [CrossRef]

36. Fu, H.Z.; Li, X.F.; Liu, W.F.; Ma, Y.M.; Gao, T.; Hong, X.H. Electronic and dynamical properties of NiAl studied from first principles. *Intermetallics* **2011**, *19*, 1959–1967. [CrossRef]

37. Reed, R.P.; Clark, A.F. *American Society of Metals*; Metals Park: Geauga County, OH, USA, 1983.

38. Friák, M.; Šob, M.; Vitek, V. Ab initio calculation of tensile strength in iron. *Philos. Mag.* **2003**, *83*, 3529–3537. [CrossRef]

39. Fu, H.Z.; Peng, W.M.; Gao, T. Structural and elastic properties of ZrC under high pressure. *Mater. Chem. Phys.* **2009**, *115*, 789–794. [CrossRef]

40. Pettifor, D.G. Theoretical predictions of structure and related properties of intermetallics. *Mater. Sci. Technol.* **1992**, *8*, 345–349. [CrossRef]

41. Johnson, R.A. Analytic nearest-neighbor model for fcc metals. *Phys. Rev. B* **1988**, *37*, 3924–3931. [CrossRef]

Article

The Immobilization of Oxindole Derivatives with Use of Cube Rhombellane Homeomorphs

Przemysław Czeleń * and Beata Szefler

Department of Physical Chemistry, Faculty of Pharmacy, Collegium Medicum, Nicolaus Copernicus University, Kurpinskiego 5, 85-096 Bydgoszcz, Poland
* Correspondence: przemekcz@cm.umk.pl

Received: 24 June 2019; Accepted: 8 July 2019; Published: 10 July 2019

Abstract: A key aspect of modern drug research is the development of delivery methods that ensure the possibility of implementing targeted therapy for a specific biological target. The use of nanocarriers enables to achieve this objective, also allowing to reduce the toxicity of used substances and often extending their bioavailability. Through the application of docking methods, the possibility of using cube rhombellanes as potential carriers for two oxindole derivatives was analyzed. In the studies, compounds identified as inhibitors of the CDK2 enzyme and a set of nanostructures proposed by the Topo Cluj Group were used. The popular fullerene molecule C_{60} was used as the reference system. The estimated binding affinities and structures of obtained complexes show that use of functionalized cube rhombellanes containing hydrogen bond donors and acceptors in their external molecular shell significantly increases ligand affinity toward considered nanocariers, compared to classic fullerenes. The presented values also allow to state that an important factor determining the mutual affinity of the tested ligands and nanostructures is the symmetry of the analyzed nanocarriers and its influence on the distribution of binding groups (aromatic systems, donors and acceptors of hydrogen bonds) on the surface of nanoparticles.

Keywords: oxindole derivatives; docking; rhombellanes; fullerenes

1. Introduction

In recent medicinal chemistry one of the most important research goals is to find ways for effective drug delivery. In numerous works the subject matter was taken up using natural substances such as albumin [1], chitosan [2,3], gelatin [4], or synthetic compounds as gold complexes [5,6], hydrogels [7], magnetic iron oxides [8] or polymers [9–13], as one of the drug carriers, however, one of the most dynamically developing groups of compounds are nanoparticles based on fullerene compounds [14–18]. Application of considered drug carriers not only improves the way of their delivery to the biological target, but also increases the bioavailability and significantly reduces the occurrence of side effects associated with the toxicity of substances with pharmacological potential [15,19]. In the case of many new nanocarriers, a significant extension in the time of drug bioactivity is observed [20], often also related with an increase in the observed pharmacological effect [21]. Among the compounds belonging to the fullerenes group, numerous studies have focused on C_{60} molecule which exhibits biological activity towards cells [14]. Its chemical structure enables for penetration of cell membrane [22–24], however in low concentrations it is non-toxic for living organisms [15,16,25,26]. The characteristic spherical surface of this nanomolecule created by conjugated carbon rings allows for π-stacking interactions with biological targets containing aromatic systems like aminoacids, vitamins, nucleic acid bases or drugs [27]. The numerous research show that C_{60} is a good carrier for drugs containing extended aromatic clusters in their structure like doxorubicin used in chemotherapy [18], however, there is a large group of compounds which chemical structure does not provide such significant possibilities in terms of staking impacts. The presence of numerous donors and acceptors of hydrogen bonds in such compounds indicates on the second potential stabilizing factor, which can occur through the selection of an appropriate nanocarrier. Such group of compounds encompasses for example oxindole derivatives which are well known as competitive inhibitors of many enzymes included in the group of kinases. The two compounds presented in Figure 1 namely CHEMBL272026 (Lig1) [28] and CHEMBL410072 (Lig2) [29] represent such group of compounds. This molecules were identified as inhibitors of Cyclin dependent kinase 2 (cdk2) and potential drugs in therapy of cancer diseases [29–34]. In the chemical structure of these compounds, except for the oxindole core containing two condensed rings, many donors and acceptors of hydrogen bond can be found which are localized in the molecule core and both side chains of considered molecules. Such chemical structure of immobilized compounds requires appropriate nanostructures ensuring the possibility of occurrence of both types of impacts, namely π-stacking and hydrogen bonds. The presented requirements are met by the structures proposed by Diudea [35,36], namely the cube rhombellanes and functionalized structures obtained by addition of second layer to the molecule core connected by ester or amide bonds. The Cube rhombellane (Cube-rbl) homeomorphs (functionalized systems) contain a core created by a system of cyclic molecules connected by ether bonds, e.g., hydroxyl derivatives of cyclohexane or benzene and examples of such structures are presented in Figure 2. Often in such systems the hyper-adamantane motif is found (1, 6) created by system of tri- and di-hydroxy benzene derivatives, however in many systems an internal core is created only by cyklohexane derivatives, like for 156 molecule (2) and 308a4 core (4). The Cube-rbl homeomorphs contain also external layer created by addition of eight subunits connected by three bonds of different type, e.g., ester for 308 (3) and 396 (7) or amide 372 (5) see Figure 2, the chemical definition of all Cube-rbl homeomorphs used in this work is presented in Table 1.

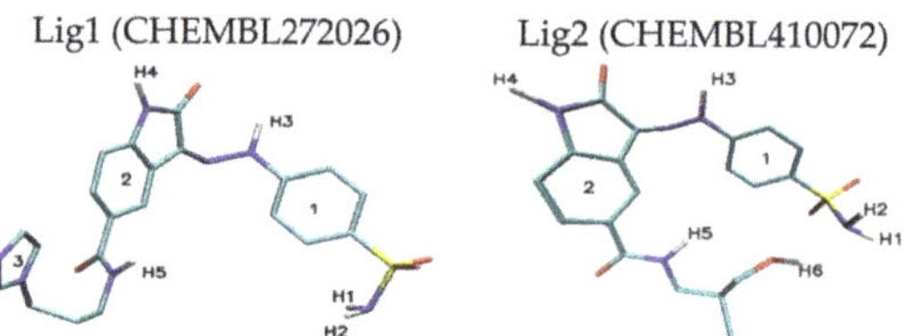

Figure 1. Graphic representation of Lig1 and Lig2 molecules. The following colors are assigned to the chemical elements: azure—carbon; dark blue—nitrogen; red—oxygen; white—hydrogen; yellow—sulphur.

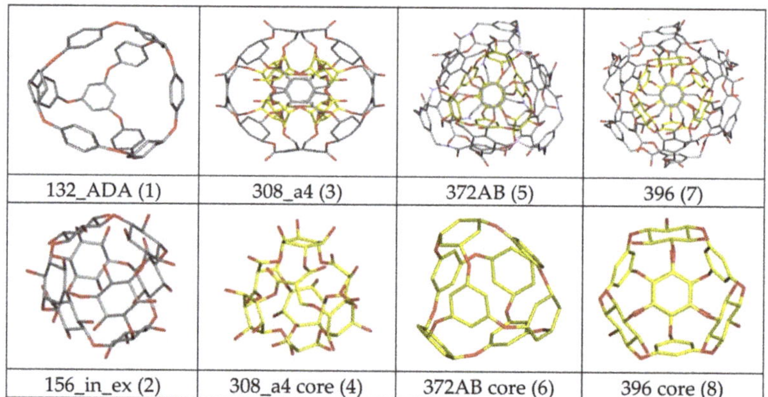

Figure 2. Graphic representation of the cube rhombellanes (1,2) and functionalized systems (3,5,7). Systems presented in yellow colour (4,6,7) correspond to the molecular cores of functionalized systems.

Table 1. Description of cube rhombellanes (Core) and functionalized structures with external layer (C-rbl). Value chem. type define chemical bond type characteristic for specific layer of considered molecule (core—molecular core; ex. shell—external layer of the nanomolecule).

Nanostructure Name	Structure Type	Chem.Type Core/ex. Shell	Atoms Quantity				
			All	C	H	O	N
144 ex_ex/in_ex	Core	Ether	144	48	84	24	0
156 ex_ex/in_ex	Core	Ether	156	48	84	36	0
308 a4/b4	C-rbl	Ether/Ester	308	100	124	84	0
360 a/b	C-rbl	Ether/Ester	360	168	108	84	0
372AB	C-rbl	Ether/Amide; Ester	372	180	120	60	12
396	C-rbl	Ether/Ester	396	192	132	72	0
420	C-rbl	Ether/Amide	420	192	156	48	24
444	C-rbl	Ether/Amide	444	192	180	48	24
456	C-rbl	Ether/Amide	456	192	180	60	24
ADA_132	ADA/rbl	Ether	132	60	60	12	0

2. Methods

The structures of ligand molecules used during docking stage, namely CHEMBL272026 ((3Z)-*N*-[3-(1H-Imidazol-1-yl)propyl]-2-oxo-3-[(4-sulfamoylphenyl)hydrazono]-5-indolinecarboxamide) (Lig1) and CHEMBL410072 ((3Z)-*N*-(3-Hydroxy-2,2-dimethylpropyl)-2-oxo-3-[(4-sulfamoylphenyl) hydrazono]-5-indolinecarboxamide) (Lig2), were obtained from CHEMBL Database [37]. The structures of nanomolecules, namely Cube Rhombellane homeomorphs, were made by Topo Cluj Group [35,36], the C_{60} structure was downloaded from Brookhaven Protein Database PDB [38]. The docking procedure was realized with use of AutoDockVina [39]. All structures used during docking stage, specifically ligands and nanoparticles, contain only polar hydrogen atoms which can participate in hydrogen bond creation. For all considered nanocarriers there were established the grid box dimensions equal to $26 \times 26 \times 26$ Å and the chosen values ensure free interactions of ligand molecules with each fragment of nanostructure surface. All molecules used during docking procedure were processed with use of AutoDock Tools package [40]. During calculations the scoring function was applied with exhaustiveness parameter equal 20, further increase of this value did not contribute to the increase of the reproducibility of the results obtained for calculations realized with different values of random seed. The structural analysis of obtained complexes, related with identification of hydrogen bonds and distances between aromatic systems, were realized with use of VMD package [41].

3. Results and Discussion

The affinities of considered oxindole derivatives namely Lig1 and Lig2 toward chosen nanostructures were estimated with use of docking procedure. All obtained values are presented in Tables 2 and 3. Both considered ligand structures exhibit similar affinity relative to reference structure fullerene C60, expressed by values equal to −5.6 and −5.3 kcal/mol. The observed difference in presented values correlates with the number of cyclic systems interacting with fullerene surfaces. All ligands conformations obtained for this nanostructure are characterized by high energetic and structural similarity, what confirms slight differences of binding affinity values (ΔGmax/ΔGmin/ΔGavrage) characterizing considered population of analyzed complexes. The values obtained for the C_{60} structure were used to assess the suitability of other nanostructures as nanocarriers for the considered ligands. In the case of both used inhibitors, it was noticed that seven nanostructures exhibit similar or significantly higher affinity compared to the reference system. Among all complexes formed by Lig1, the systems containing 360a, 372AB, 308a4, 308b4 and ADA_132 nanocarriers deserve special attention. For all mentioned complexes there is observed a noticeable increase of binding affinity values and the best manifestation of this phenomenon are the differences in values of binding constant the increase of which is in the range from 40 to 175% Table 2 relative to reference system. The complexes of C_{60} fullerene with molecules exhibiting pharmacological potential are mainly stabilized by stacking interactions between aromatic systems of both molecules. In the case of Lig1 molecule its structure contains three cyclic systems which exhibit affinity relative to the aromatic surface of fullerene.

The graphic representation Figure 3 and values presented in Table 4 confirm that all cyclic systems of ligand molecule preserve planar orientation relative to C_{60} fullerene, also relatively small distances oscillating around 3.6 Å indicate on a dominant role of stacking interactions in complex stabilization. The non-homoatomic structure of the rest of nanostructures used during docking stage provides a wider range of impacts including the potential of hydrogen bonds formation. The data presented in Table 4 unambiguously show that such type of interactions plays an important role in stabilization of complexes obtained during docking procedure. The complex of Lig1 molecule and 360a nanomolecule, characterized by highest values of binding affinity (−6.20 kcal/mol), is stabilized by four hydrogen bonds involving all hydrogen bond donors present in the considered molecule, their lengths (from 2.07 to 2.73 Å) indicate on medium or weak strength of such interactions. The ligand conformation also supports the presence of stacking interactions between aromatic systems of oxindole core of molecule and fullerene nanocarrier Figure 3, which primarily confirms their mutual planar orientation and relatively short distance equal to 3.68 Å. In the case of the rest of chosen complexes, there is observed a much higher share of stacking interactions involved in their structure stabilization, which manifests through the involvement of at least two cyclic systems in interactions with rhombllane structures. The conformations of Lig1 molecules characterized by planar or quasi planar orientation of two rings distant from nanocarriers in the range from 3.41 to 4.43, caused a much smaller activity in creation of hydrogen bonds by considered molecule, what confirms not only their quantity but also measured distances, placed near the border limit of hydrogen bond length. Such situation can be observed even in the case of the second complex (372AB) characterized by binding affinity lower only by 0.1 kcal/mol relative to 360a system. The important factor describing the quality of binding with nanocarrier is the difference between energy values characterizing all system conformations obtained during docking stage. In the case of best structure, the difference between best and worst conformation does not exceed 0.3 [kcal/mol] (0.18 relative to the average), while in the case of the second system this difference increases to 0.6 [kcal/mol] (0.43 relative to the average). In the case of Lig2 molecule, the systems formed with 360a, 308a4, 308b4, 372AB and 396 nanocarriers are of particular interest. All mentioned systems exhibit large increase of binding constant relative to reference system placed in the range from 132 to 440%. The highest affinity, analogically like in the case of first group of complexes, was found for 360a molecule. All complexes of this nanomolecule obtained during docking stage, regardless of the ligand molecule conformation, are characterized by significantly higher energy values compared to the rest of considered systems (ΔG_{max} = −6.30; $\Delta G_{AVERAGE}$ = −6.08 kcal/mol). The Lig2 inhibitor interacts with this nanocarrier through hydrogen bonds and stacking interactions. The data presented in Table 4 shows that four hydrogen atoms from Lig2 molecule are involved in creation of bonds and two of them exhibit the possibility of interaction with two different hydrogen bond acceptors. The presented distances, placed in the range from 2.09 to 2.48 Å, allow to classify them as medium and weak strength interactions. The Lig2 conformation presented in Figure 4 confirms the planar orientation of aromatic ring from oxindole core relative to the aromatic system of 360a subunit. The quite short distance between cyclic systems equals 3.66 Å and theirs mutual orientation indicates an important share of stacking interaction in complex stability. In the case of Lig2 complexes with the other four nanostructures, all best obtained systems are characterized by similar affinity value equal to −5.80 kcal/mol However, the analysis of minimal and average values presented in Table 3 allows to conclude that population of conformations obtained for 308b4 molecule is characterized by the highest cohesion of energy values. Such observation could point to better affinity of the Lig2 ligand to the surface of this nanostructure. All four complexes are stabilized by stacking interactions of one of aromatic systems, in the case of 308a4 and 396 systems from the oxindole core of molecule, while in the case of 308b4 and 372AB from the side chain. The quantity and quality of hydrogen bonds created by ligand molecule in complexes is also quite similar, however some discrepancies are observed. The Lig2 complex with 308b4 molecule is maintained by four medium strength impacts, while in the case of 396 and 372AB there are observed three and two hydrogen bonds for 308a4. In the case of 396 and 308a4 systems, there was found also one potential impact slightly exceeding border limit of hydrogen bond.

Table 2. Values of binding affinity [kcal/mol] of Lig1 molecule relative to nanosystems obtained during docking stage.

Nanostructure Name	ΔG [kcal/mol]				Binding Constant [K_{max}]	Difference of Kmax Relative to C_{60} [%]
	MAX	**MIN**	**AVERAGE**	**SD**		
144_ex_ex	−3.50	−3.20	−3.30	0.12	367.73	−97.11
144_in_ex	−3.90	−3.70	−3.81	0.09	722.33	−94.33
156_ex_ex	−4.10	−3.70	−3.88	0.11	1012.37	−92.05
156_in_ex	−4.20	−3.80	−3.92	0.14	1198.51	−90.59
308a4	−6.00	−5.60	−5.76	0.13	25,006.81	96.43
308b4	−5.90	−5.50	−5.74	0.14	21,123.08	65.92
360a	−6.20	−5.90	−6.02	0.11	35,047.76	175.30
360b	−5.50	−5.10	−5.26	0.11	10,753.59	−15.53
372AB	−6.10	−5.50	−5.67	0.21	29,604.61	132.54
396	−5.60	−5.20	−5.37	0.13	12,730.76	0.00
420	−5.70	−5.20	−5.42	0.15	15,071.46	18.39
444	−5.30	−5.00	−5.12	0.10	7672.76	−39.73
456	−5.50	−5.00	−5.26	0.17	10,753.59	−15.53
ADA_132	−5.80	−5.50	−5.58	0.10	17,842.53	40.15
C_{60}	−5.60	−5.40	−5.50	0.07	12,730.76	—

Table 3. Values of binding affinity [kcal/mol] of Lig2 molecule relative to nanosystems obtained during docking stage.

Nanostructure Name	ΔG [kcal/mol]				Binding Constant [K_{max}]	Difference of Kmax Relative to C_{60} [%]
	MAX	**MIN**	**AVERAGE**	**SD**		
144_ex_ex	−3.60	−3.50	−3.57	0.05	435.35	−94.33
144_in_ex	−4.00	−3.70	−3.80	0.10	855.14	−88.85
156_ex_ex	−3.90	−3.70	−3.79	0.06	722.33	−90.59
156_in_ex	−4.20	−3.90	−4.03	0.09	1198.51	−84.38
308a4	−5.80	−5.50	−5.59	0.11	17,842.53	132.54
308b4	−5.80	−5.60	−5.72	0.08	17,842.53	132.54
360a	−6.30	−5.80	−6.08	0.15	41,491.70	440.77
360b	−5.10	−4.80	−4.92	0.13	5474.56	−28.65
372AB	−5.80	−5.40	−5.58	0.13	17,842.53	132.54
396	−5.80	−5.50	−5.68	0.11	17,842.53	132.54
420	−5.60	−5.40	−5.49	0.08	12,730.76	65.92
444	−5.00	−4.70	−4.83	0.11	4624.33	−39.73
456	−5.00	−4.80	−4.91	0.08	4624.33	−39.73
ADA_132	−5.40	−5.10	−5.19	0.11	9083.48	18.39
C_{60}	−5.30	−5.20	−5.21	0.03	7672.76	—

Table 4. Distances characterizing interactions involved in stabilization of Lig1 and Lig2 complexes.

	Lig1					
	308b4	**308a4**	**360a**	**372AB**	**ADA 132**	**C_{60}**
H1/H2	2.44	2.50/2.85	2.42	—	—	—
H3	3.03	—	2.52	2.97	—	—
H4	—	2.97	2.73	—	—	—
H5	2.04	3.05	2.07	2.38	2.81	—
Stacking 1/2/3	3.74/—/3.96	—/3.74/4.34	—/3.68/—	3.55/3.95/—	3.94; 3.41/—/3.79	3.66/3.59/3.63
	Lig2					
	308b4	**308a4**	**360a**	**372AB**	**396**	**C_{60}**
H1/H2	1.92	1.96	2.09/2.32	2.52	1.95	—
H3	2.46	—	2.47	—	2.33/2.50	—
H4	—	3.02	—	—	—	—
H5	2.36	—	2.42	2.32	3.15	—
H6	2.21	2.07/2.80	2.09/2.48	2.03	2.09	—
Stacking 1/2	3.59/—	—/3.59	—/3.66	3.67/—	—/3.44	3.55/3.60

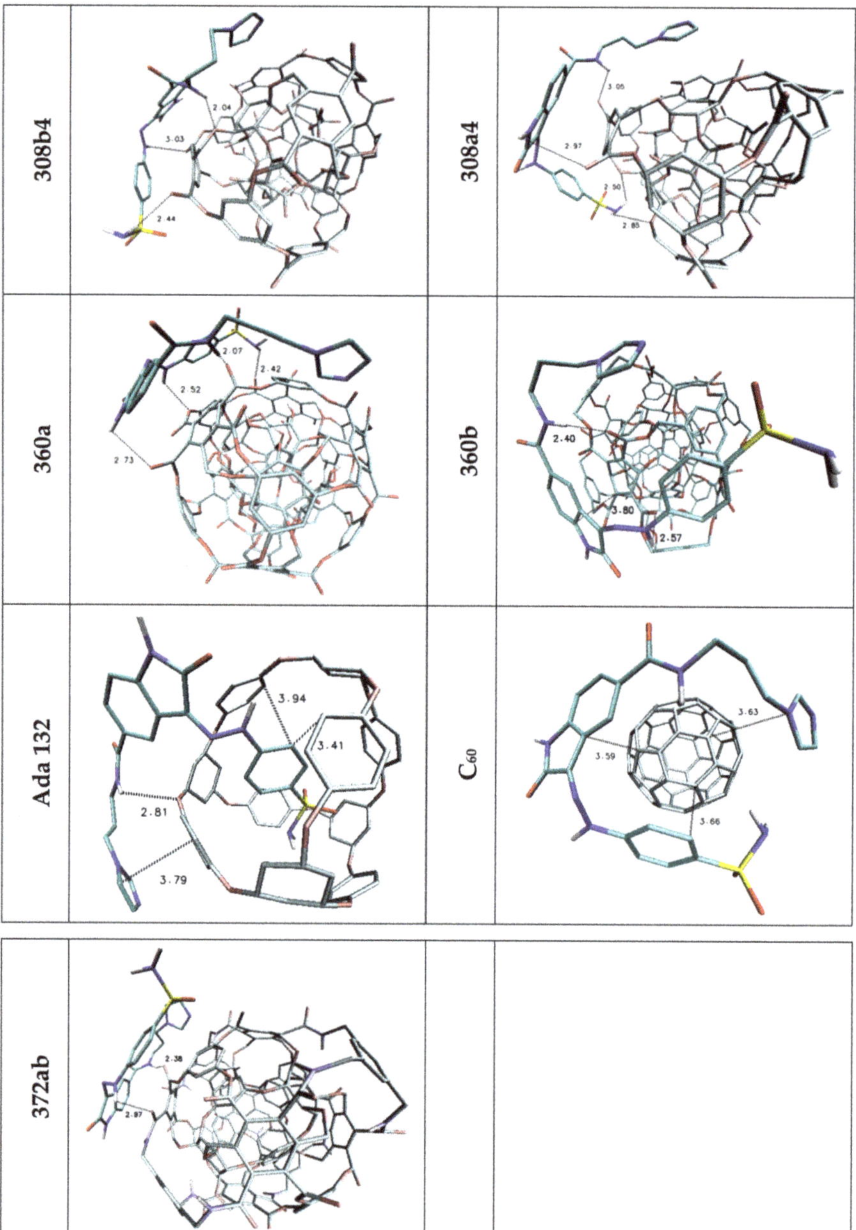

Figure 3. Graphic representations of Lig1 complexes with nanomolecules characterized by highest values od binding enthalpy. The following colors are assigned to the chemical elements: azure—carbon; dark blue—nitrogen; red—oxygen; white—hydrogen; yellow—sulphur.

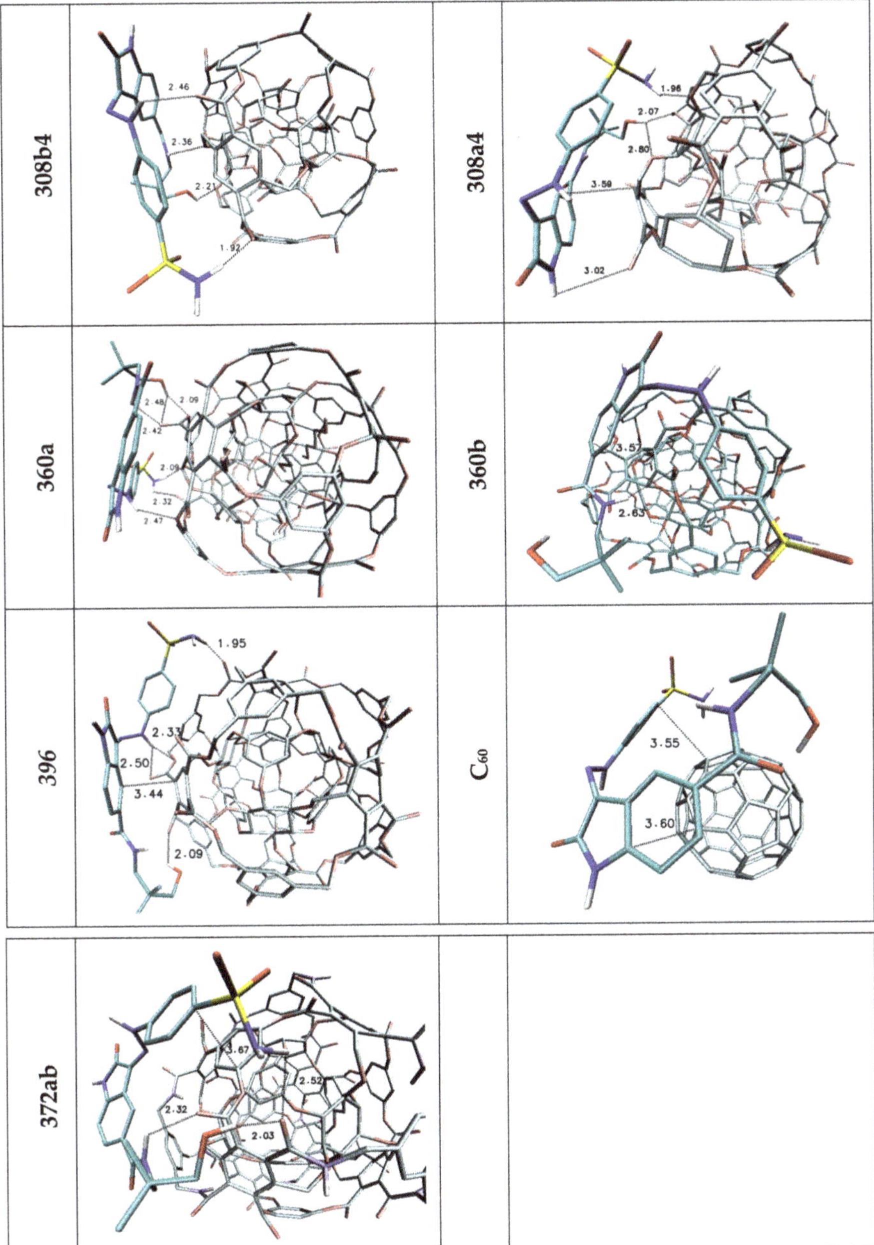

Figure 4. Graphic representations of Lig2 complexes with nanomolecules characterized by highest values od binding entalphy. The following colors are assigned to the chemical elements: azure—carbon; dark blue—nitrogen; red—oxygen; white—hydrogen; yellow—Sulphur.

The Binding Affinity and Conformational Diversity

The cube rhombellanes and its homeomorphs exihibit conformational diversity strictly related with structure of molecule core and its symmetry obtained by use of different structural isomers of cyclohexane or changes of dihedral angles describing its mutual orientation. In Figure 5 there are presented exemplary structures of nanosystems exhibiting properties of structural isomers. The molecule 144 has two structural isomers ("ex_ex" and "in_ex"), i.e., the change of one of cyclohexane form causes the occurrence of two nanostructures exhibiting different symmetry what can be observed in Figure 5a–d. The example of structural diversity of more complex molecules are systems 360 a/b, in this case the different mutual orientation of benzene and cyclohexane rings causes the structural change of the molecular core (Figure 5f,h), which also contributed to the change of the topology of external shell of considered molecules (Figure 5e,g). Tables 1 and 2 contain also data representing ligand affinities toward different structural conformers of considered nanocarriers. Among all considered nanostructures, two different structural forms related with changes of molecular symmetry may be found in the case 144, 156, 308 and 360 molecules. Only in one case both considered structural forms of the nanocarrier exhibit similar binding affinity towards considered ligand molecules, namely both complexes of 308 isomers are characterized by similar maximal, minimal and average values of ΔG and observed differences do not exceed −0.13 kcal/mol. Another situation is observed for conformers 144, 156 and 360. In the case of the first two molecules different symmetry (see Figure 5a–d) caused slight but noticeable changes of their affinities towards both ligands, observed differences in values are placed in the range from 0.3 to 0.5 kcal/mol and for both systems "in_ex" form is preferred. The most significant impact of structural changes related with symmetry of central core of nanomolecule is observed in the case of 360 system. The structural diversities of compared systems presented in Figure 5e,h significantly affected binding properties of these two nanocarriers toward considered ligands. All values characterizing these systems, including maximal, average and minimal binding affinities, unambiguously show belter binding properties of 360a structure and the observed differences from 0.7 to 1.2 kcal/mol indicate a low binding efficiency of ligands with the 306b structure, lower even than in the reference system C_{60}.

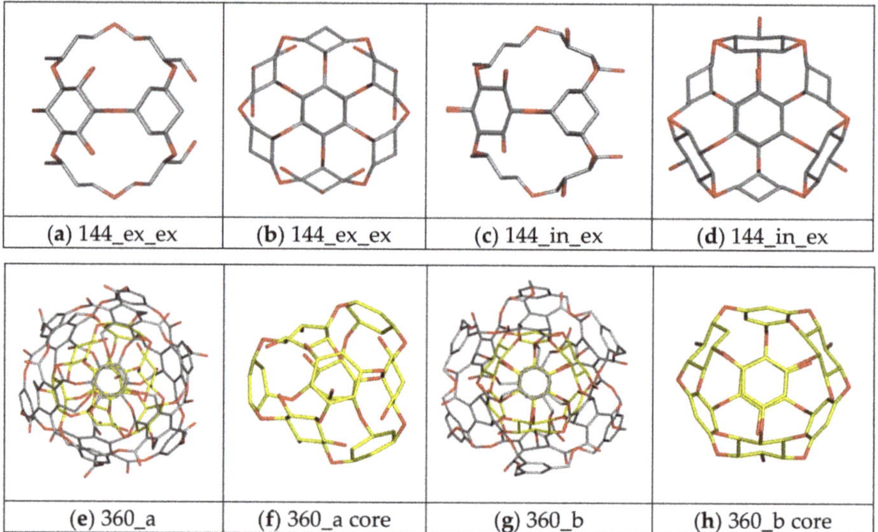

(a) 144_ex_ex	**(b)** 144_ex_ex	**(c)** 144_in_ex	**(d)** 144_in_ex
(e) 360_a	**(f)** 360_a core	**(g)** 360_b	**(h)** 360_b core

Figure 5. Graphic representation of different symmetry of chosen nanomolecules used during docking stage. Systems presented in yellow color correspond to the different structural isomers of the molecular core of the 360 particle.

4. Conclusions

The ligands molecules considered in this work contain aromatic systems, hydrogen bond donors and acceptors. Such chemical structures indicate that the most appropriate nanocarrier should use both types of potential impacts during creation of nanocomplexes. Presented data show that properly designed nanomolecules can improve binding of molecules with pharmacological potential relative to commonly used fullerenes. Through the use of benzene or cyclohexane rings linked by ester and amide bonds, such structures of nanoparticles can be obtained that provide better affinity to the analyzed inhibitors than in the case of other fullerene systems. The confirmation of this are the complexes of both analyzed ligands with 360a molecule, characterized by up to 4–5 fold higher value of binding constant than in the case of reference structure, fullerene C_{60}. However, not always the presence of hydrogen bond donors and acceptors in nanomolecule surface contributes to an increase of binding affinity and durability of complexes, because an equally important factor determining these features is the symmetry of the analyzed nanosystems defining the mutual distribution of aromaticity centers, as well as hydrogen donors and acceptors on the surface of the molecule. The most extreme example of such dependence are the isomers of the 360 molecule, i.e., the "a" form of this molecule proved to be the most effective in interacting with the tested ligands (−6.2/6.3 kcal/mol), while the "b" form showed worse properties than the referred fullerene (−5.5/−5.1 kcal/mol).

Author Contributions: Conceptualization, P.C.; methodology, P.C.; software, P.C.; validation, P.C.; formal analysis, P.C.; investigation, P.C.; resources, P.C. and B.S.; data curation, P.C.; writing-original draft preparation, P.C.; writing-review & editing, P.C. and B.S.; visualization, P.C.; supervision, P.C.; project administration, P.C.; funding acquisition, P.C.

Funding: This research received no external funding.

Acknowledgments: Thanks to the Professor MV Diudea for sharing the structures. This research was supported by PL-Grid Infrastructure (http://www.plgrid.pl/en).

Conflicts of Interest: The authors declare that there are no conflicts of interest.

References

1. Damascelli, B.; Patelli, G.L.; Lanocita, R.; Di Tolla, G.; Frigerio, L.F.; Marchianò, A.; Garbagnati, F.; Spreafico, C.; Tichà, V.; Gladin, C.R.; et al. A Novel Intraarterial Chemotherapy Using Paclitaxel in Albumin Nanoparticles to Treat Advanced Squamous Cell Carcinoma of the Tongue: Preliminary Findings. *Am. J. Roentgenol.* **2003**, *181*, 253–260. [CrossRef] [PubMed]

2. Huang, M.; Khor, E.; Lim, L.-Y. Uptake and cytotoxicity of chitosan molecules and nanoparticles: Effects of molecular weight and degree of deacetylation. *Pharm. Res.* **2004**, *21*, 344–353. [CrossRef] [PubMed]

3. Dyer, A.M.; Hinchcliffe, M.; Watts, P.; Castile, J.; Jabbal-Gill, I.; Nankervis, R.; Smith, A.; Illum, L. Nasal delivery of insulin using novel chitosan based formulations: A comparative study in two animal models between simple chitosan formulations and chitosan nanoparticles. *Pharm. Res.* **2002**, *19*, 998–1008. [CrossRef] [PubMed]

4. Cascone, M.G.; Lazzeri, L.; Carmignani, C.; Zhu, Z. Gelatin nanoparticles produced by a simple W/O emulsion as delivery system for methotrexate. *J. Mater. Sci. Mater. Med.* **2002**, *13*, 523–526. [CrossRef] [PubMed]

5. Paciotti, G.F.; Myer, L.; Weinreich, D.; Goia, D.; Pavel, N.; McLaughlin, R.E.; Tamarkin, L. Colloidal gold: A novel nanoparticle vector for tumor directed drug delivery. *Drug Deliv.* **2004**, *11*, 169–183. [CrossRef]

6. Hainfeld, J.F.; Slatkin, D.N.; Smilowitz, H.M. The use of gold nanoparticles to enhance radiotherapy in mice. *Phys. Med. Biol.* **2004**, *49*, N309–N315. [CrossRef] [PubMed]

7. Gupta, M.; Gupta, A.K. Hydrogel pullulan nanoparticles encapsulating pBUDLacZ plasmid as an efficient gene delivery carrier. *J. Control. Release* **2004**, *99*, 157–166. [CrossRef] [PubMed]

8. Gupta, A.K.; Gupta, M. Synthesis and surface engineering of iron oxide nanoparticles for biomedical applications. *Biomaterials* **2005**, *26*, 3995–4021. [CrossRef] [PubMed]

9. Kim, S.Y.; Lee, Y.M.; Baik, D.J.; Kang, J.S. Toxic characteristics of methoxy poly(ethylene glycol)/poly(epsilon-caprolactone) nanospheres; in vitro and in vivo studies in the normal mice. *Biomaterials* **2003**, *24*, 55–63. [CrossRef]

10. Alyautdin, R.N.; Petrov, V.E.; Langer, K.; Berthold, A.; Kharkevich, D.A.; Kreuter, J. Delivery of loperamide across the blood-brain barrier with polysorbate 80-coated poly(butylcyanoacrylate) nanoparticles. *Pharm. Res.* **1997**, *14*, 325–328. [CrossRef]

11. Kreuter, J.; Ramge, P.; Petrov, V.; Hamm, S.; Gelperina, S.E.; Engelhardt, B.; Alyautdin, R.; von Briesen, H.; Begley, D.J. Direct evidence that polysorbate-80-coated poly(butylcyanoacrylate) nanoparticles deliver drugs to the CNS via specific mechanisms requiring prior binding of drug to the nanoparticles. *Pharm. Res.* **2003**, *20*, 409–416. [CrossRef] [PubMed]

12. Panyam, J.; Zhou, W.Z.; Prabha, S.; Sahoo, S.K.; Labhasetwar, V. Rapid endo-lysosomal escape of poly(DL-lactide-*co*-glycolide) nanoparticles: Implications for drug and gene delivery. *FASEB J.* **2002**, *16*, 1217–1226. [CrossRef] [PubMed]

13. Weissenböck, A.; Wirth, M.; Gabor, F. WGA-grafted PLGA-nanospheres: Preparation and association with Caco-2 single cells. *J. Control. Release* **2004**, *99*, 383–392. [CrossRef] [PubMed]

14. Cataldo, F.; Da Ros, T. *Medicinal Chemistry and Pharmacological Potential of Fullerenes and Carbon Nanotubes*; Springer: Berlin, Germany, 2008; ISBN 9781402068454.

15. De Jong, W.H.; Borm, P.J.A. Drug delivery and nanoparticles: Applications and hazards. *Int. J. Nanomed.* **2008**, *3*, 133–149. [CrossRef]

16. Andrievsky, G.; Klochkov, V.; Derevyanchenko, L. Is the C_{60} Fullerene Molecule Toxic?! *Fuller. Nanotub. Carbon Nanostructures* **2005**, *13*, 363–376. [CrossRef]

17. Szefler, B. Nanotechnology, from quantum mechanical calculations up to drug delivery. *Int. J. Nanomed.* **2018**, *13*, 6143–6176. [CrossRef] [PubMed]

18. Panchuk, R.R.; Prylutska, S.V.; Chumakl, V.V.; Skorokhyd, N.R.; Lehka, L.V.; Evstigneev, M.P.; Prylutskyy, Y.I.; Berger, W.; Heffeter, P.; Scharff, P.; et al. Application of C60 Fullerene-Doxorubicin Complex for Tumor Cell Treatment In Vitro and In Vivo. *J. Biomed. Nanotechnol.* **2015**, *11*, 1139–1152. [CrossRef]

19. Morgen, M.; Bloom, C.; Beyerinck, R.; Bello, A.; Song, W.; Wilkinson, K.; Steenwyk, R.; Shamblin, S. Polymeric Nanoparticles for Increased Oral Bioavailability and Rapid Absorption Using Celecoxib as a Model of a Low-Solubility, High-Permeability Drug. *Pharm. Res.* **2012**, *29*, 427–440. [CrossRef]

20. Gao, Z.; Zhang, L.; Sun, Y. Nanotechnology applied to overcome tumor drug resistance. *J. Control. Release* **2012**, *162*, 45–55. [CrossRef]

21. Turov, V.V.; Chehun, V.F.; Barvinchenko, V.N.; Krupskaya, T.V.; Prylutskyy, Y.I.; Scharff, P.; Ritter, U. Low-temperature 1H-NMR spectroscopic study of doxorubicin influence on the hydrated properties of nanosilica modified by DNA. *J. Mater. Sci. Mater. Med.* **2011**, *22*, 525–532. [CrossRef]

22. Schuetze, C.; Ritter, U.; Scharff, P.; Fernekorn, U.; Prylutska, S.; Bychko, A.; Rybalchenko, V.; Prylutskyy, Y. Interaction of *N*-fluorescein-5-isothiocyanate pyrrolidine-C_{60} with a bimolecular lipid model membrane. *Mater. Sci. Eng. C* **2011**, *31*, 1148–1150. [CrossRef]

23. Qiao, R.; Roberts, A.P.; Mount, A.S.; Klaine, S.J.; Ke, P.C. Translocation of C60 and Its Derivatives Across a Lipid Bilayer. *Nano Lett.* **2007**, *7*, 614–619. [CrossRef] [PubMed]

24. Prylutska, S.; Bilyy, R.; Overchuk, M.; Bychko, A.; Andreichenko, K.; Stoika, R.; Rybalchenko, V.; Prylutskyy, Y.; Tsierkezos, N.G.; Ritter, U. Water-soluble pristine fullerenes C_{60} increase the specific conductivity and capacity of lipid model membrane and form the channels in cellular plasma membrane. *J. Biomed. Nanotechnol.* **2012**, *8*, 522–527. [CrossRef] [PubMed]

25. Prylutska, S.V.; Grynyuk, I.I.; Grebinyk, S.M.; Matyshevska, O.P.; Prylutskyy, Y.I.; Ritter, U.; Siegmund, C.; Scharff, P. Comparative study of biological action of fullerenes C_{60} and carbon nanotubes in thymus cells. *Materwiss. Werksttech.* **2009**, *40*, 238–241. [CrossRef]

26. Johnston, H.J.; Hutchison, G.R.; Christensen, F.M.; Aschberger, K.; Stone, V. The Biological Mechanisms and Physicochemical Characteristics Responsible for Driving Fullerene Toxicity. *Toxicol. Sci.* **2010**, *114*, 162–182. [CrossRef]

27. Evstigneev, M.P.; Buchelnikov, A.S.; Voronin, D.P.; Rubin, Y.V.; Belous, L.F.; Prylutskyy, Y.I.; Ritter, U. Complexation of C_{60} Fullerene with Aromatic Drugs. *Chem Phys Chem* **2013**, *14*, 568–578. [CrossRef] [PubMed]

28. Czeleń, P. Molecular dynamics study on inhibition mechanism of CDK-2 and GSK-3β by CHEMBL272026 molecule. *Struct. Chem.* **2016**, *27*, 1807–1818. [CrossRef]

29. Czeleń, P. Inhibition mechanism of CDK-2 and GSK-3β by a sulfamoylphenyl derivative of indoline—A molecular dynamics study. *J. Mol. Model.* **2017**, *23*, 230. [CrossRef]

30. Besson, A.; Dowdy, S.F.; Roberts, J.M. CDK Inhibitors: Cell Cycle Regulators and Beyond. *Dev. Cell* **2008**, *14*, 159–169. [CrossRef] [PubMed]

31. Canavese, M.; Santo, L.; Raje, N. Cyclin dependent kinases in cancer. *Cancer Biol. Ther.* **2012**, *13*, 451–457. [CrossRef] [PubMed]
32. Child, E.S.; Hendrychov, T.; McCague, K.; Futreal, A.; Otyepka, M.; Mann, D.J. A cancer-derived mutation in the PSTAIRE helix of cyclin-dependent kinase 2 alters the stability of cyclin binding. *Biochim. Biophys. Acta Mol. Cell Res.* **2010**, *1803*, 858–864. [CrossRef] [PubMed]
33. Malumbres, M.; Barbacid, M. Cell cycle, CDKs and cancer: A changing paradigm. *Nat. Rev. Cancer* **2009**, *9*, 153–166. [CrossRef] [PubMed]
34. Malumbres, M.; Barbacid, M. To cycle or not to cycle: A critical decision in cancer. *Nat. Rev. Cancer* **2001**, *1*, 222–231. [CrossRef] [PubMed]
35. Diudea, M.V.; Lungu, C.N.; Nagy, C.L.; Diudea, M.V.; Lungu, C.N.; Nagy, C.L. Cube-Rhombellane Related Structures: A Drug Perspective. *Molecules* **2018**, *23*, 2533. [CrossRef] [PubMed]
36. Szefler, B.; Czeleń, P.; Diudea, M. V Docking of indolizine derivatives on cube rhombellane functionalized homeomorphs. *Stud. Univ. Babes-Bolyai Chem.* **2018**, *63*, 7–18. [CrossRef]
37. ChEMBL. Available online: https://www.ebi.ac.uk/chembl/ (accessed on 1 March 2016).
38. Kim, K.H.; Ko, D.K.; Kim, Y.T.; Kim, N.H.; Paul, J.; Zhang, S.Q.; Murray, C.B.; Acharya, R.; Kim, Y.H.; DeGrado, W.F.; et al. Protein-directed self-assembly of a fullerene crystal. *Nat. Commun.* **2016**, *7*, 11429. [CrossRef] [PubMed]
39. Trott, O.; Olson, A.J. AutoDock Vina: Improving the speed and accuracy of docking with a new scoring function, efficient optimization, and multithreading. *J. Comput. Chem.* **2010**, *31*, 455–461. [CrossRef] [PubMed]
40. Bartashevich, E.V.; Potemkin, V.A.; Grishina, M.A.; Belik, A.V. A Method for Multiconformational Modeling of the Three-Dimensional Shape of a Molecule. *J. Struct. Chem.* **2002**, *43*, 1033–1039. [CrossRef]
41. Humphrey, W.; Dalke, A.; Schulten, K. VMD: Visual molecular dynamics. *J. Mol. Graph.* **1996**, *14*, 33–38. [CrossRef]

 *symmetry*

Article

Docking of Cisplatin on Fullerene Derivatives and Some Cube Rhombellane Functionalized Homeomorphs

Beata Szefler * and Przemysław Czeleń

Department of Physical Chemistry, Faculty of Pharmacy, Collegium Medicum, Nicolaus Copernicus University, Kurpińskiego 5, 85-096, Bydgoszcz, Poland
* Correspondence: beatas@cm.umk.pl

Received: 12 June 2019; Accepted: 27 June 2019; Published: 3 July 2019

Abstract: Cisplatin (cisPt) is one of the strongest anticancer agents with proven clinical activity against a wide range of solid tumors. Its mode of action has been linked to its ability to crosslink with the canonical purine bases, primarily with guanine. Theoretical studies performed at the molecular level suggest that such nonspecific interactions can also take place with many competitive compounds, such as vitamins of the B group, containing aromatic rings with lone-pair orbitals. This might be an indicator of reduction of the anticancer therapeutic effects of the Cisplatin drug in the presence of vitamins of the B group inside the cell nucleus. That is why it seems to be important to connect CisPt with nanostructures and in this way prevent the drug from combining with the B vitamins. As a proposal for a new nanodrug, an attempt was made to implement Cispaltin (CisPt) ligand on functionalized C_{60} fullerenes and on a cube rhombellane homeomorphic surface. The symmetry of the analyzed nanostructures is an important factor determining the mutual affinity of the tested ligand and nanocarriers. The behavior of Cisplatin with respect to rhombellane homeomorphs and functionalized fullerenes C_{60}, in terms of their (interacting) energy, geometry and topology was studied and a detailed analysis of structural properties after docking showed many interesting features.

Keywords: cube rhombellane homeomorph; functionalized fullerene C_{60}; Cisplatin (CisPt), nanostructure; molecular docking; affinity

---

1. Introduction

Cisplatinum (cisPt) (Figure 1) is one of the strongest anticancer agents with proven clinical activity against a wide range of solid tumors [1–5]. As a chemotherapeutic drug it has been used for treatment of numerous human cancers including lung, bladder, testicular, head, and neck cancers [1] and ovarian carcinomas [4]. It is effective against various types of cancers, including carcinomas, lymphomas, germ cell tumors and sarcomas [2]. Its mode of action has been linked to its ability to crosslink with the canonical purine bases, primarily with guanine, and to a lesser extent with adenine, found within double helical DNA. This in turn seriously interferes with DNA repair mechanisms, causing DNA damage and subsequently inducing apoptosis [6] in cancer cells.

Figure 1. Cisplatinum (cisPt; [Pt(NH$_3$)$_2$Cl$_2$]) [5].

However, theoretical studies performed at the molecular level suggest that such nonspecific interactions can also take place with many competitive compounds, such as vitamins containing aromatic rings with lone-pair orbitals [7]. It is argued that such direct analogy to interactions with canonical purines can impair the therapeutic effect of Cisplatin. This might be an indicator of reduction of the anticancer therapeutic effects of the Cisplatin drug in the presence of vitamins of the B group inside the cell nucleus [7].

That is why it seems to be important to connect CisPt with nanostructures and in this way make it prevent the drug from combining with the B vitamins. After passing through the cellular barrier, a rupture of the complex would occur and the free cisPt could interact with bases in the interior of the cell (Figure 2).

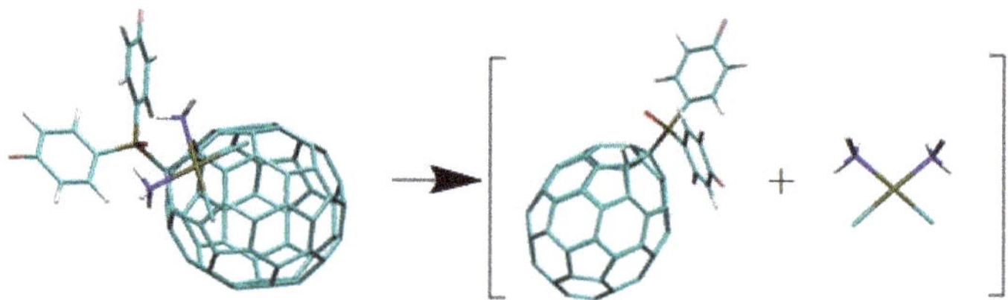

Figure 2. Rupture of the complex after passing through the cellular barrier.

Rhombellans are new structures defined by Diudea [8–13] with rings in the form of rhombs or squares, which could be an alternative to classical carrier nanostructures. This new class of compounds, having drug-like properties, and creating bio-nano devices, can be suitable for medical chemistry.

Rhombellanes have certain specific features: (i) all strong rings are squares/rhombs; (ii) vertex classes consist of only non-connected vertices; (iii) the Omega polynomial has a single term: $1X^{\wedge}|E|$; (iv) a line graph of the parent graph has a Hamiltonian circuit; and (v) they contain at least one $K_{2.3}$ complete bipartite subgraph or the smallest rhombellane Rbl.5.

Rhombellanes may be synthesized as real molecules, which represent a new class of hypothetical structures. Quantum calculations (at the B3LYP/6-31G (d, p) level of theory) [14,15] support the hypothesis that these substructures (at every level of complexity) are energetically feasible in the hope of a real synthesis [16–20].

There are many types of these structures, among which only those have been used which have functional groups that allow attachment of CisPt (Figure 3).

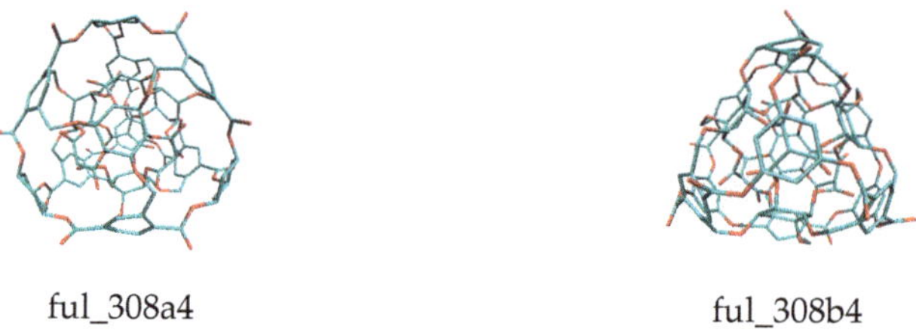

ful_308a4   ful_308b4

Figure 3. *Cont.*

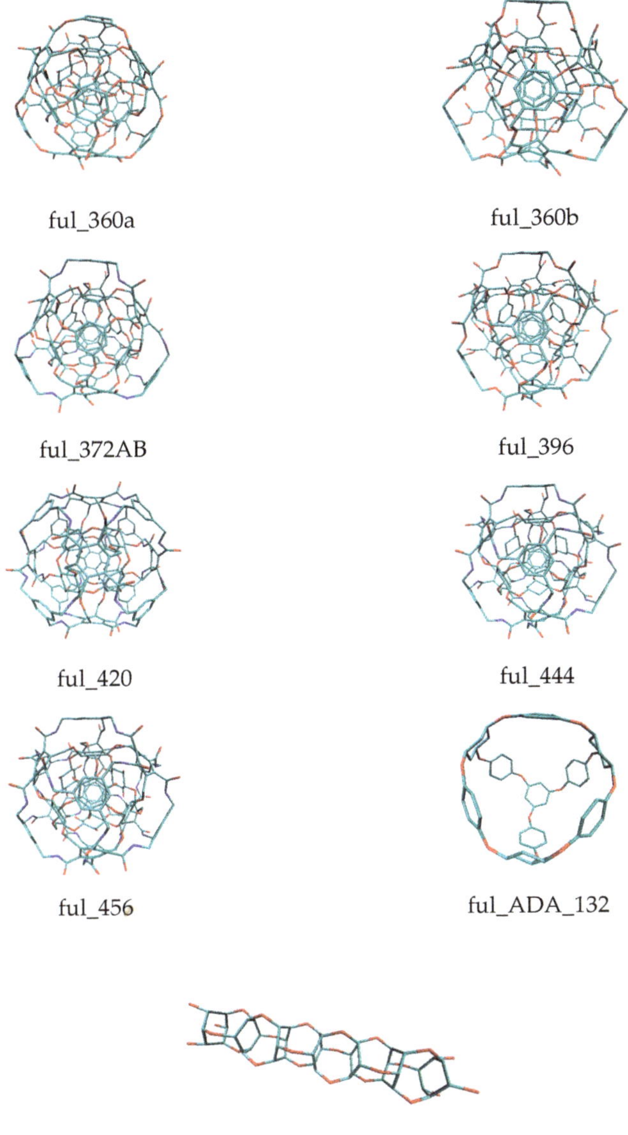

Figure 3. Graphic representation of the cube rhombellanes.

Despite the initially expected chemical inactivity of C_{60} and its derivatives, it was found that fullerenes can be functionalized [21]. Due to the mode of functionalization, there were distinguished exo- and endohedral forms of fullerenes and heterofullerenes. The essence of exohedral chemistry is the chemical reactions of attachment "outside" of fullerene molecules, in which the structure of the carbon cage remains unchanged. Unlike other organic, aliphatic and aromatic compounds, fullerenes do not contain hydrogen atoms or other functional groups, so they cannot undergo substitution reactions (except for Heterofullerens). Substitution reactions occur only with "functionalized" fullerenes, i.e. when they have been attached to specific groups of atoms. C_{60} and its homologues have interesting and often unique properties, and after functionalization the C_{60} molecule can bind any functional group (Figure 4) [21]. At the same time, it is indifferent, non-toxic and so small that it easily comes into contact with cells, proteins and viruses. In the docking procedure with Cisplatin several exohedral forms of C_{60} fullerenes were used, such as 1-(Diethoxyphosphorylmethyl)-7-(2,2,2-trifluoroethoxy)(C_{60}-Ih)[5,6]fullerene;9-(diethoxyphosphorylmethyl)-1H-(C_{60}-Ih)[5,6]fullerene;

Methyl 9-(2-trimethylsilylethyl)(C_{60}-Ih)[5,6]fullerene-1-carboxylate and others (Figure 4).

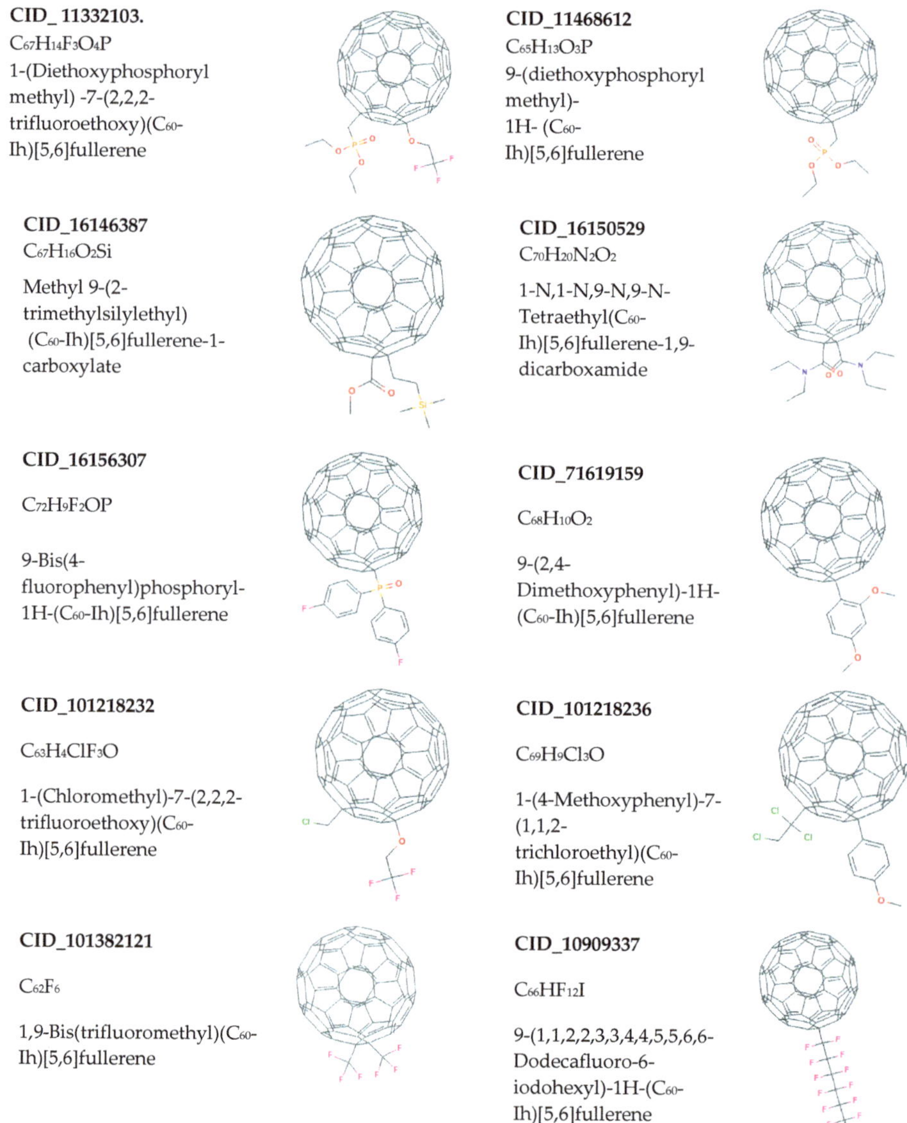

CID_ 11332103.
$C_{67}H_{14}F_3O_4P$
1-(Diethoxyphosphoryl methyl) -7-(2,2,2-trifluoroethoxy)(C_{60}-Ih)[5,6]fullerene

CID_11468612
$C_{65}H_{13}O_3P$
9-(diethoxyphosphoryl methyl)-1H- (C_{60}-Ih)[5,6]fullerene

CID_16146387
$C_{67}H_{16}O_2Si$
Methyl 9-(2-trimethylsilylethyl) (C_{60}-Ih)[5,6]fullerene-1-carboxylate

CID_16150529
$C_{70}H_{20}N_2O_2$
1-N,1-N,9-N,9-N-Tetraethyl(C_{60}-Ih)[5,6]fullerene-1,9-dicarboxamide

CID_16156307
$C_{72}H_9F_2OP$
9-Bis(4-fluorophenyl)phosphoryl-1H-(C_{60}-Ih)[5,6]fullerene

CID_71619159
$C_{68}H_{10}O_2$
9-(2,4-Dimethoxyphenyl)-1H-(C_{60}-Ih)[5,6]fullerene

CID_101218232
$C_{63}H_4ClF_3O$
1-(Chloromethyl)-7-(2,2,2-trifluoroethoxy)(C_{60}-Ih)[5,6]fullerene

CID_101218236
$C_{69}H_9Cl_3O$
1-(4-Methoxyphenyl)-7-(1,1,2-trichloroethyl)(C_{60}-Ih)[5,6]fullerene

CID_101382121
$C_{62}F_6$
1,9-Bis(trifluoromethyl)(C_{60}-Ih)[5,6]fullerene

CID_10909337
$C_{66}HF_{12}I$
9-(1,1,2,2,3,3,4,4,5,5,6,6-Dodecafluoro-6-iodohexyl)-1H-(C_{60}-Ih)[5,6]fullerene

Figure 4. Graphic representation of functionalized fullerenes C_{60}.

For the transposition of cisPt at the cellular level through numerous barriers and to avoid their connection with the B group vitamins [7], an attempt was made to deposit Cisplatin on rhombellanes and functionalized fullerenes C_{60}, as possible nanodrug complexes. The symmetry of the analyzed nanostructures is an important factor determining the mutual affinity of the tested ligand and nanocarriers.

Detailed analysis of structural properties after docking showed many interesting features. Behavior of Cisplatin with respect to rhombellane homeomorphs and functionalized fullerenes C_{60}, in terms of their (interacting) energy, geometry and topology, was studied. After the docking procedure, the optimal values of ligand-rhombellane and ligand- C_{60} fullerene affinities were found, which is an important result for homeomorphs.

2. Methods

2.1. Docking Procedure

The structures of rhombellane homeomorphs were prepared by Topo Cluj Group [8–13], while the functionalized C_{60} structures were downloaded from the PubChem Database [22]. The docking procedure was realized with the use of non-commercial docking program AutoDock 4.2 [23,24]. This employs a stochastic Lamarckian genetic algorithm for computing ligand conformations and simultaneously minimizing its scoring function, which approximates the thermodynamic stability of the ligand bound to the target fullerene. For all considered nanocarriers there were established grid box dimensions equal to $26 \times 26 \times 26$ Å. The center of the grid box was placed in the center of the considered nanomolecules. In most cases, the center of the grid box had xyz coordinates equal to 000. During the docking procedure, the molecules were loaded and stored as pdb-files, after assigning hydrogen bonds. The investigated ligands were loaded, and their torsions along the rotatable bonds were assigned and then saved as "ligand.pdbqt". The grid menu after loading "pdbqt" was toggled [25]. For the search of the ligand–Rbl nanostructure and ligand-C_{60} interactions, the map files were selected directly, setting up the grid points separately for each structure. The Lamarckian genetic algorithm completed the docking parameter files [26]. The structural analysis of the obtained complexes related with the identification of hydrogen bonds was realized with use of the Visual Molecular Dynamics (VMD) package [27].

2.2. Results and Discussion

The results are presented in the following tables and figures. Rhombellane structures are given by their atom number.

Table 1 and Figure 5 represent binding affinity in kcal/mol of the ligand cisPt molecule relative to spherical nanosystems, such as rhombellane structures (first eleven structures in tables) and functionalized fullerene C_{60} (structures 12 to 21 in tables) obtained during docking stage.

Table 1. Values of binding affinity [kcal/mol] of the ligand CisPt molecule relative to spherical nanosystems (rhombellane structures and functionalized fullerene C_{60}) obtained during the docking stage, generated by software Autodock 4.2.

	Binding Energy (kcal/mol)									
360b	−2.66	−2.6	−2.61	−2.49	−2.45	−2.42	−2.33	−2.44	−2.25	−2.42
372	−2.51	−2.38	−2.36	−2.31	−2.25	−2.25	−2.3	−2.2	−2.19	−2.17
396	−2.34	−2.29	−2.28	−2.19	−2.23	−2.23	−2.22	−2.22	−2.16	−2.13
420	−2.17	−2.16	−2.16	−2.16	−2.16	−2.16	−2.16	−2.09	−2.09	−2.09
444	−2.72	−2.71	−2.72	−2.69	−2.67	−2.67	−2.65	−2.63	−2.63	−2.63
456	−2.62	−2.62	−2.56	−2.56	−2.56	−2.55	−2.52	−2.45	−2.43	−2.41
ADA132	−2.3	−2.23	−2.27	−2.26	−2.23	−2.05	−2.16	−2.16	−2.04	−1.98
308a4	−2.16	−2.11	−1.97	−2.16	−2.14	−2.14	−2.14	−2.04	−1.95	−1.84
308b4	−2.09	−1.9	−2.08	−2.01	−1.98	−2.05	−1.94	−1.89	−1.84	−1.8
360a	−2.35	−2.23	−2.23	−2.17	−2.11	−2.22	−2.19	−2.17	−2.09	−2.02

Table 1. *Cont.*

	Binding Energy (kcal/mol)									
stf114	−1.9	−1.88	−1.87	−1.85	−1.78	−1.83	−1.76	−1.74	−1.82	−1.67
CID_11332103	−2.47	−2.47	−2.41	−2.41	−2.4	−2.4	−2.4	−2.37	−2.35	−2.19
CID_11468612	−2.54	−2.53	−2.51	−2.51	−2.5	−2.49	−2.48	−2.47	−2.47	−2.46
CID_16146387	−2.1	−2.09	−2.08	−2.08	−2.06	−2.01	−2	−2.06	−2.05	−2.04
CID_16150529	−2.14	−2.1	−2.09	−2.06	−2.04	−2.02	−1.98	−1.98	−1.97	−1.97
CID_16156307	−3.44	−3.41	−3.39	−3.37	−3.37	−3.37	−3.36	−3.36	−3.35	−3.33
CID_71619159	−1.74	−1.67	−1.73	−1.72	−1.57	−1.57	−1.56	−1.53	−1.49	−1.48
CID_101218232	−2.83	−2.82	−2.77	−2.76	−2.71	−2.63	−2.48	−2.46	−2.45	−2.4
CID_101218236	−3.13	−3.09	−3.07	−3.06	−3.05	−2.98	−2.93	−2.92	−2.91	−2.79
CID_101382121	−0.82	−0.8	−0.78	−0.78	−0.75	−0.8	−0.8	−0.77	−0.74	−0.73
CID_10909337_C	−0.97	−0.96	−0.96	−0.96	−0.96	−0.95	−0.94	−0.94	−0.94	−0.93

Among Rbl homeomorphs, Cisplatin has the highest affinity for Rbl 444 with an affinity value of −2.72 kcal/mol and for Rbl 360b with an affinity value of −2.66 kcal / mol (Table 1, Figure 5). The lowest binding energy is observed in the case of Staffanes (stf) 114 (Table 1, Figure 5). Among functionalized structures of C_{60} fullerene, the best affinity is shown by CID_16156307, followed by CID_10121832 and finally the CID_10121831 structure with values of affinity equal to −3.44, −3.13 and −2.83 kcal/mol, respectively (Table 1, Figure 5).

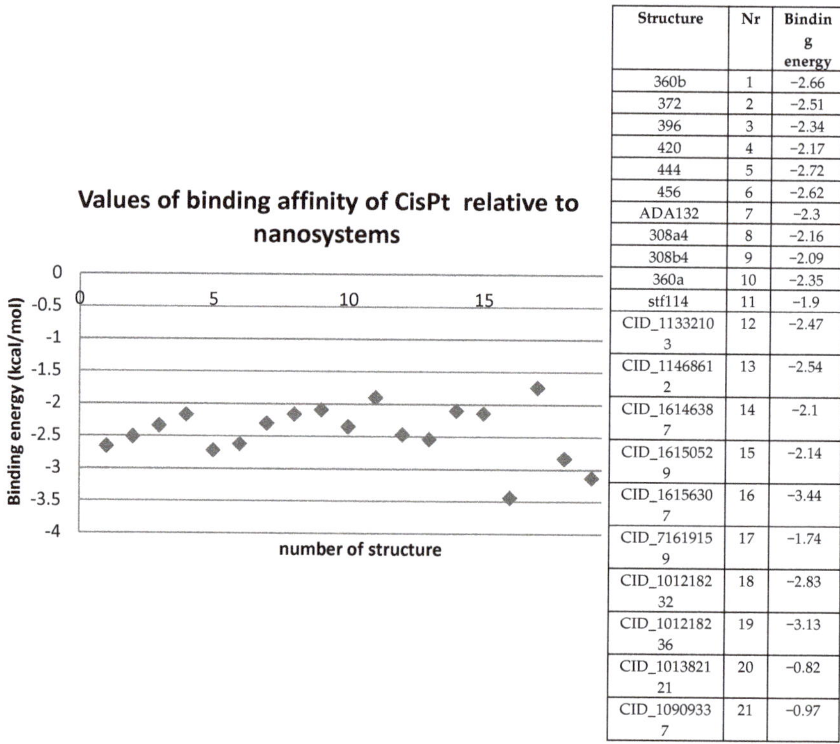

Structure	Nr	Binding energy
360b	1	−2.66
372	2	−2.51
396	3	−2.34
420	4	−2.17
444	5	−2.72
456	6	−2.62
ADA132	7	−2.3
308a4	8	−2.16
308b4	9	−2.09
360a	10	−2.35
stf114	11	−1.9
CID_11332103	12	−2.47
CID_11468612	13	−2.54
CID_16146387	14	−2.1
CID_16150529	15	−2.14
CID_16156307	16	−3.44
CID_71619159	17	−1.74
CID_101218232	18	−2.83
CID_101218236	19	−3.13
CID_101382121	20	−0.82
CID_10909337	21	−0.97

Figure 5. Graphical presentation of the values of binding affinity [kcal/mol] of the ligand CisPt molecule relative to spherical nanosystems (rhombellane structures and functionalized fullerene C_{60}) obtained during docking stage, generated by software Autodock 4.2.

In general, the examined structures show a significantly larger affinity in the case of C_{60} functionalized fullerene than in the case of Rbl homeomorphs. The obtained affinity values are collected in Tables 2 and 3.

Table 2. The best binding affinity of ligand CisPt, maximum and minimum bind energy, and Kmax values of the binding constant estimated with use of the binding free energy obtained for the best complex of ligand with nanostructure Rbl after the docking procedure, generated by software Autodock 4.2.

Name of Nanostructure	Maximum Binding Energy	Minimum Binding Energy	Average	SD	Binding Constant [K_{max}]
360b	−2.66	−2.25	−2.47	0.12	89.1
372	−2.51	−2.17	−2.29	0.10	69.2
396	−2.34	−2.13	−2.23	0.06	51.9
420	−2.17	−2.09	−2.14	0.03	39.0
444	−2.72	−2.63	−2.67	0.03	98.6
456	−2.62	−2.41	−2.53	0.07	83.3
ADA132	−2.30	−1.98	−2.17	0.10	48.5
308a4	−2.16	−1.84	−2.07	0.11	38.3
308b4	−2.09	−1.80	−1.96	0.10	34.0
360a	−2.35	−2.02	−2.18	0.09	52.8
stf114	−1.90	−1.67	−1.81	0.07	24.7

The values of binding energy are correlated with values of the K_{max} constant and the highest values are obtained in the case of 444 fullerene and 360b nanostructure (Table 2).

The two last columns show the equilibrium K value of the bonds, calculated using the equation:

$$K_B = exp^{\left(-\frac{\Delta G_b}{RT}\right)}$$

where K_b is the binding constant, R is the gas constant (J/mol*K), T is the temperature of 298 K and ΔG_b is the binding affinity (J/mol).

The higher the K value, the more the reaction proceeds towards the formation of the complex.

Table 3. The best binding affinity of the ligand CisPt, maximum and minimum binding energy and Kmax values of the binding constant estimated with use of the binding free energy obtained for the best complex of the ligand with functionalized fullerene C_{60} after the docking procedure, generated by software Autodock 4.2.

Name of Nanostructure	Maximum Binding Energy	Minimum Binding Energy	Average	SD	Binding Constant [K_{max}]	Type
CID_11332103	−2.47	−2.19	−2.39	0.07	64.6	$C_{67}H_{14}F_3O_4P$
CID_11468612	−2.54	−2.46	−2.50	0.03	72.8	$C_{65}H_{13}O_3P$
CID_16146387	−2.10	−2.00	−2.06	0.03	34.6	$C_{67}H_{16}O_2Si$
CID_16150529	−2.14	−1.97	−2.04	0.06	37.0	$C_{70}H_{20}N_2O_2$
CID_16156307	−3.44	−3.33	−3.38	0.03	332.3	$C_{72}H_9F_2OP$
CID_71619159	−1.74	−1.48	−1.61	0.09	18.9	$C_{68}H_{10}O_2$
CID_101218232	−2.83	−2.40	−2.63	0.16	118.7	$C_{63}H_4ClF_3O$
CID_101218236	−3.13	−2.79	−2.99	0.10	196.9	$C_{69}H_9Cl_3O$
CID_101382121	−0.82	−0.73	−0.78	0.03	4.0	$C_{62}F_6$
CID_10909337	−0.97	−0.93	−0.95	0.01	5.1	$C_{66}HF_{12}I$

Table 3 presents K_{max} values of the binding constant estimated with use of the binding free energy obtained for the best complex of the ligand with functionalized fullerene C_{60}. The highest values of the binding energy are also here correlated with values of the constant K_{max}. This value is higher in the case of derivatives CID_16156307 of C_{60} fullerene, followed by CID_101218236 and CID_101218232 (Table 3).

Cisplatin (cisPt) and functionalized fullerene C_{60} easily create two or three hydrogen bonds between them with strong and medium strength (Figure 6).

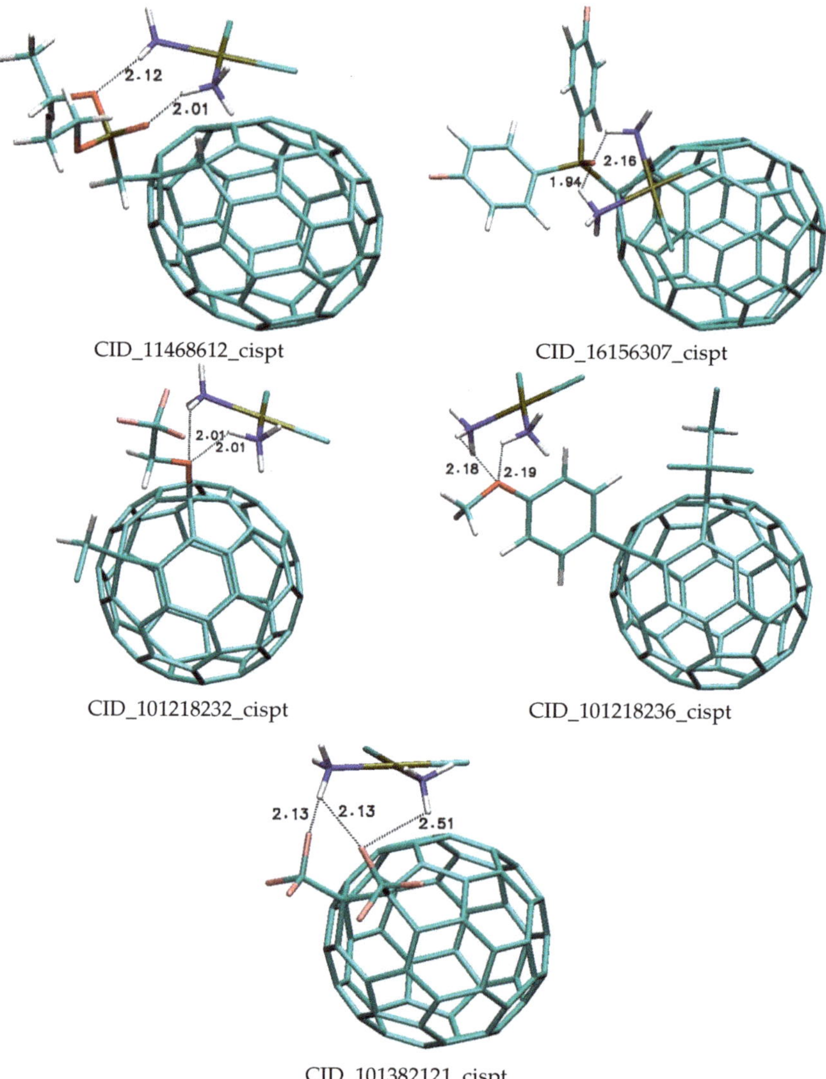

Figure 6. The structure of functionalized fullerene C_{60} and Cisplatin complexes.

The criterion for classification of the strength of hydrogen bonds is the assessment of the distance between acceptor and hydrogen atoms: strong interactions are characterized by a distance <1.6 Å, medium by values in the range from 1.6 Å to 2.0 Å, and weak by distances <3 Å.

The complex nanostructure CID_16156307 ($C_{72}H_9F_2OP$)–CisPt has been created by two hydrogen bonds with medium strengths, i.e., 1.94 Å and 2.16 Å; both are between the oxygen of the phosphoryl group of the nanocarrier and the hydrogen of the amino groups of cisPt (Figure 6). Again, two hydrogen bonds have been created in the case of CID_ 101218236 ($C_{69}H_9Cl_3O$) and CID_101218232 ($C_{63}H_4ClF_3O$) complexes. The first case involves the oxygen atom of the methoxy group while the second involves the oxygen atom of the ethoxy group, and both form a Hydrogen Bond (HB) with the hydrogen atom of the amino groups of cisPt with values 2.18 Å, 2.19 Å and 2.01 Å (Figure 6). Three hydrogen bonds appeared only in the case of CID_10138212 ($C_{62}F_6$), all with medium strength, i.e., 2.13 Å, 2,13 Å and 2.51 Å (Figure 6), respectively. The structures of formed complexes are presented in Figure 6.

Despite the fact that affinity is higher in the case of functionalized fullerene C_{60}–cisPt, compared to affinities of Rbl–CisPt complexes, the latter created a larger number of hydrogen bonds (Figure 7). The best binding energy has been observed for the 444 nanostructure with four hydrogen bonds of medium strength, i.e., 1.88 Å, 1.95 Å, 1.96 Å and 2.03 Å, respectively. Also, for the 360b nanostructure there were created four hydrogen bonds with medium strength, i.e., 2.08 Å, 2.09 Å, 2.26 Å and 2.56 Å, respectively (Figure 7). The worst affinity is represented by the stf114 fullerene. Even so, in this case there are also four hydrogen bonds formed between the nanostructure and Cisplatin. The structures of formed complexes are presented in Figure 7.

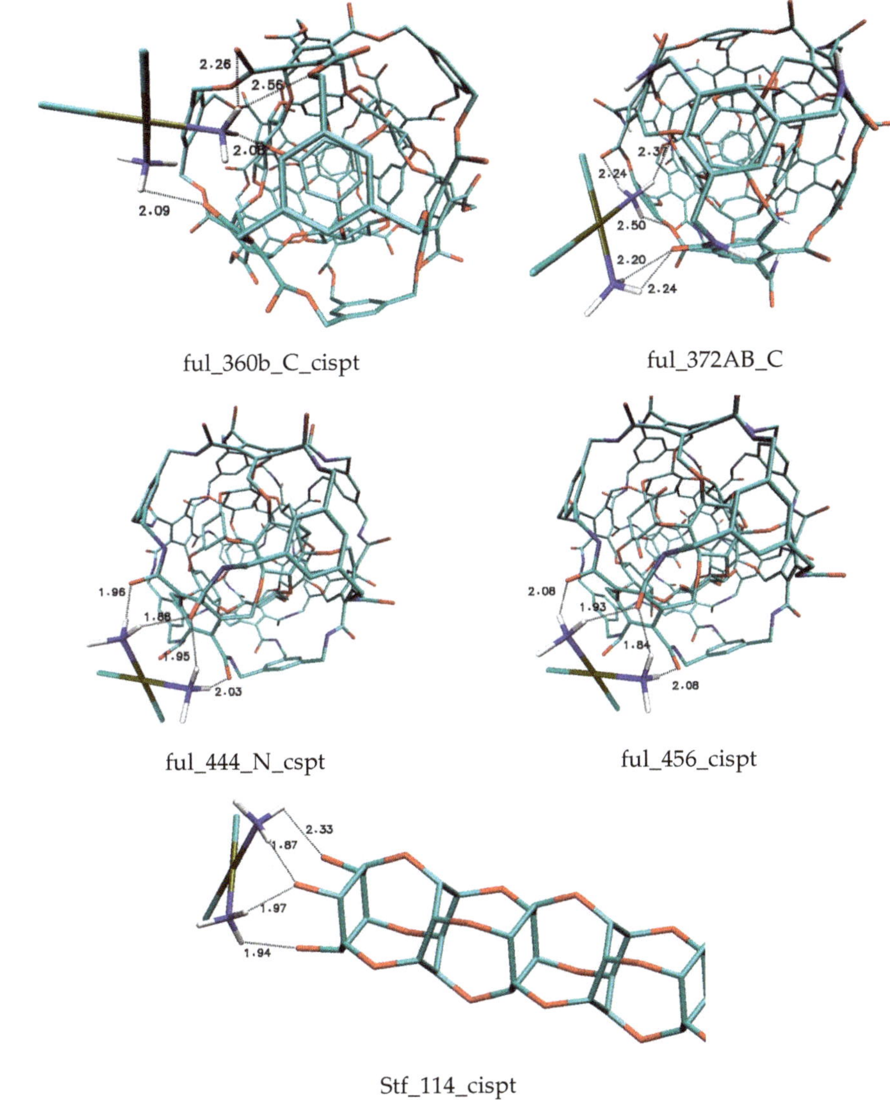

Figure 7. The structure of cube rhombellane and Cisplatin complexes.

3. Conclusions

As a proposal for a new nanodrug, an attempt was made to implement the Cisplatin (cisPt) ligand on the functionalized C_{60} fullerenes and on a cube rhombellane homeomorphic surface. Cisplatin (cisPt) is one of the strongest anticancer agents with proven clinical activity against a wide range of solid tumors. Theoretical studies suggest that nonspecific interactions of Cisplatin can also take place with many competitive compounds, such as vitamins of the B group, containing aromatic rings with lone-pair orbitals. It is argued that such a direct analogy to the interaction of canonical purines can impair the therapeutic effect of Cisplatin, which might be an indicator of the reduction of the anticancer therapeutic effects of the Cisplatin drug in the presence of vitamins of the B group inside the cell nucleus. This is why it seems to be important to connect cisPt Should be "cisPt" with nanostructures and, in this way, make it impossible to combine the drug with the B vitamins. Behavior of Cisplatin with respect to rhombellane homeomorphs and functionalized fullerenes C_{60}, in terms of their (interacting) energy, geometry and topology was studied. The symmetry of the analyzed nanostructures is an important factor determining the mutual affinity of the tested ligand and nanocarriers. The docking procedure was realized with use of AutoDock 4.2. Detailed analysis of structural properties after docking showed many interesting features. Among Rbl homeomorphs, Cisplatin has the highest affinity for Rbl 444 with an affinity value of −2.72 kcal/mol and Rbl 360b with an affinity value of −2.66 kcal/mol. Among functionalized structures of C_{60} fullerene the best affinity is shown by CID_16156307, followed by CID_10121832 and finally CID_10121831, with values of affinity equal to −3.44, −3.13 and −2.83 kcal/mol, respectively. In general, the examined structures show a significantly larger affinity in the case of C_{60} functionalized fullerene than in the case of Rbl homeomorphs. Cisplatin (cisPt) and functionalized fullerene C_{60} easily create two or three hydrogen bonds between them with strong and medium strength. High ligand-nanostructure affinity is reflected by the number and most of all the quality of formed hydrogen bonds. However, despite the fact that affinity is better in the case of functionalized fullerene C_{60}–cisPt compared with affinities of complexes Rbl–CisPt Should be "cisPt", the latter created a larger number of hydrogen bonds but of lesser quality. The performed investigations enabled the identification of the most promising rhombellane and functionalized C_{60} fullerene structures that could be used as nanocarriers for cisPt molecules.

Author Contributions: Conceptualization, B.S.; Methodology, B.S. and P.C.; Validation, B.S. and P.C.; Formal Analysis, B.S.; Investigation, B.S.; Resources, B.S.; Data Curation, B.S.; Writing–Original Draft Preparation, B.S.; Writing–Review & Editing, B.S. and P.C.; Visualization, B.S. and P.C.; Supervision, B.S.; Project Administration, B.S.; Funding Acquisition, B.S.

Acknowledgments: Gratitude is expressed for fruitful cooperation for many years to Professor MV Diudea; this article was supported by PL-Grid Infrastructure (http://www.plgrid.pl/en).

Conflicts of Interest: The authors declare no conflict of interest.

References

1. Frezza, M.; Hindo, S.; Chen, D.; Davenport, A.; Schmitt, S.; Tomco, D.; Dou, Q.P. Novel metals and metal complexes as platforms for cancer therapy. *Curr. Pharm. Des.* **2010**, *16*, 1813–1825. [CrossRef] [PubMed]

2. Desoize, B.; Madoulet, C. Particular aspects of platinum compounds used at present in cancer treatment. *Crit. Rev. Oncol. Hematol.* **2002**, *42*, 317–325. [CrossRef]

3. Fraval, H.N.; Rawlings, C.J.; Roberts, J.J. Increased sensitivity of UV-repair-deficient human cells to DNA bound platinum products which unlike thymine dimers are not recognized by an endonuclease extracted from micrococcus luteus. *Mutat. Res.* **1978**, *51*, 121–132. [CrossRef]

4. Wiernik, P.H.; Yeap, B.; Vogl, S.E.; Kaplan, B.H.; Comis, R.L.; Falkson, G.; Davis, T.E.; Fazzini, E.; Cheuvart, B.; Horton, J. Hexamethylmelamine and low or moderate dose cisplatin with or without pyridoxine for treatment of advanced ovarian carcinoma: A study of the eastern cooperative oncology group. *Cancer Invest.* **1992**, *10*, 1–9. [CrossRef] [PubMed]

5. Wiltshaw, E.; Kroner, T. Phase II study of cis-dichlorodiammineplatinum[II] (NSC-119875] in advanced adenocarcinoma of the ovary. *Cancer Treat. Rep.* **1976**, *60*, 55–60. [PubMed]

6.   Dasari, S.; Tchounwou, P.B. Cisplatin in cancer therapy: Molecular mechanisms of action. *Eur. J. Pharmacol.* **2014**, *740*, 364–378. [CrossRef] [PubMed]

7.   Szefler, B.; Czeleń, P.; Szczepanik, A.; Cysewski, P. Does affinity of cisplatin to B-Vitamins impair the therapeutic effect in the case of patient with lung cancer consuming carrot or beet juice? *Anticancer Agents Med Chem.* **2019**, *25*. [CrossRef] [PubMed]

8.   ISI Web of Science. 2010.

9.   Gomes, J.A.N.F.; Mallion, R.B. Aromaticity and ring currents. *Chem. Rev.* **2001**, *101*, 1349–1383. [CrossRef]

10. Cyrański, M.K.; Krygowski, T.M.; Katritzky, A.R.; Schleyer, P.v.R. To what extent can aromaticity be defined uniquely? *J. Org. Chem.* **2002**, *67*, 1333–1338. [CrossRef]

11. Chen, Z.; Wannere, C.S.; Crominboeuf, C.; Puchta, R.; Schleyer, R.v.P. nucleus-independent chemical shifts (NICS) as an aromaticity criterion. *Chem. Rev.* **2005**, *105*, 3842–3888. [CrossRef]

12. Diudea, M.V.; Lungu, C.N.; Nagy, C.L. Cube-rhombellane related structures: A drug perspective. *Molecules* **2018**, *23*, 2533. [CrossRef] [PubMed]

13. Diudea, M.V.; Pîrvan-Moldovan, A.; Pop, R.; Medeleanu, M. Medeleanu, Energy of graphs and remote graphs, in hypercubes, rhombellanes and fullerenes. *MATCH Commun. Math. Comput. Chem.* **2018**, *80*, 835–852.

14. Pauling, L.; Wheland, G.W. The nature of the chemical bond. V. The quantum mechanical calculation of the resonance energy of benzene and naphthalene and the hydrocarbon free radicals. *J. Chem. Phys.* **1933**, *1*, 362–374. [CrossRef]

15. Daudel, R.; Lefebre, R.; Moser, C. *Quantum Chemistry*; Interscience: New York, NY, USA, 1959.

16. Diudea, M.V. Rhombellanic diamond. Fullerenes, Nanotubes and Carbon. *Nanomaterials* **2018**. [CrossRef]

17. Pop, R.; Medeleanu, M.; Diudea, M.V.; Szefler, B.; Cioslowski, J. Fullerenes patched by flowers. *Cent. Eur. J. Chem.* **2013**, *11*, 527–534. [CrossRef]

18. Frisch, M.J.; Trucks, G.W.; Schlegel, H.B.; Scuseria, G.E.; Robb, M.A.; Cheeseman, J.R.; Scalmani, G.; Barone, V.; Mennucci, B.; Petersson, G.A.; et al. *Gaussian 09, Revision A.1*; Gaussian Inc.: Wallingford, CT, USA, 2009.

19. Randić, M. Aromaticity of Polycyclic Conjugated Hydrocarbons. *Chem. Rev.* **2003**, *103*, 3449–3605.

20. Diudea, M.V.; Nagy, C.L. *Periodic Nanostructures*; Springer: Dordrecht, The Netherlands, 2007.

21. Szefler, B. Nanotechnology, from quantum mechanical calculations up to drug delivery. *Int. J. Nanomed.* **2018**, *13*, 6143–6176. [CrossRef] [PubMed]

22. Pubchem. Available online: https://pubchem.ncbi.nlm.nih.gov// (accessed on 11 June 2019).

23. Morris, G.M.; Huey, R.; Lindstrom, W.; Sanner, M.F.; Belew, R.K.; Goodsell, D.S.; Olson, A.J. Autodock4 and AutoDockTools4: Automated docking with selective receptor flexibility. *J. Comput. Chem.* **2009**, *30*, 2785–2791. [CrossRef] [PubMed]

24. Shoichet, B.K.; Kuntz, I.D.; Bodian, D.L. Molecular docking using shape descriptors. *J. Comput. Chem.* **2004**, *13*, 380–397. [CrossRef]

25. Dhananjayan, K.; Kalathil, K.; Sumathy, A.; Sivanandy, P. A computational study on binding affinity of bio-flavonoids on the crystal structure of 3-hydroxy-3-methyl-glutaryl-CoA reductase—An insilico molecular docking approach. *Der Pharma Chemica.* **2014**, *6*, 378–387.

26. Abagyan, R.; Totrov, M. High-throughput docking for lead generation. *Current Opin. Chem. Biol.* **2001**, *5*, 375. [CrossRef]

27. Humphrey, W.; Dalke, A.; Schulten, K. VMD: Visual molecular dynamics. *J. Mol. Graph.* **1996**, *14*, 33–38. [CrossRef]

Article

Application of the Consonance Solvent Concept for Accurate Prediction of Buckminster Solubility in 180 Net Solvents using COSMO-RS Approach

Piotr Cysewski

Chair and Department of Physical Chemistry, Faculty of Pharmacy, Collegium Medicum of Bydgoszcz, Nicolaus Copernicus University in Toruń, Kurpińskiego 5, 85-950 Bydgoszcz, Poland; piotr.cysewski@cm.umk.pl; Tel.: +48-52-585-3611

Received: 10 June 2019; Accepted: 20 June 2019; Published: 22 June 2019

Abstract: The default COSMO-RS (Conductor like Screening Model for Real Solvents) approach is incapable of accurate computation of C60 solubility in net solvents. Additionally, there is no adequate selection of single or multiple reference solvent, which can be applied to the whole population of 180 solvents for improving prediction of mole fraction at saturated conditions. This failure cannot be addressed to inaccurate data of the Buckminster fusion, although they pose a challenge for experimental measurement due to intense sublimation of C60 at elevated temperatures and the possibility of solvates precipitation. However, taking advantage of the richness of experimental data of fullerene solubility, it is possible to identify the source of errors expressed in terms of fluidization affinity. Classification of solvents according to the value of this fluidization term allowed for formulation of a consonance solvents approach, which enables accurate prediction of C60 solubility using the single reference solvent method.

Keywords: Buckminster; solubility; COSMO-RS (Conductor like Screening Model for Real Solvents); reference solvent; net organic solvent; fusion

---

1. Introduction

Buckminster fullerene is a highly symmetric all carbon molecule, in which structure is represented by truncated icosahedron with a cage-like fused-rings made of twenty hexagons and twelve pentagons. The sixty carbon atoms placed at each vertex of each polygon are covalently bonded along each polygon edge. Fullerene C60 belongs to a broad class of Goldberg polyhedron-like carbon allotropes occurring in the form of spheres, ellipsoids or tubes. It was first generated in 1984 using a laser induced carbon vaporization in a supersonic helium beam [1]. The unique structure of C60 results [2] in its unparalleled unique physicochemical properties [3], which were very welcome in many industries taking advantage of nanomaterials as for example biomedicine [4,5], optics, electronics and cosmetics [6]. Unfortunately, low solubility of C60 in many organic solvents [7] stands for the major cost of the production from soot [8], since extraction by organic solvents is the first step for obtaining fullerenes-rich fractions further separated using HPLC [9]. Moreover, strong tendency of precipitation in the form of solvates [10–17] makes it difficult to preserve purity of the solid. This is why modeling of C60 solubility attracted so much attention and resulted in a variety of theoretical approaches, among which the best predications come so far from non-linear modeling via machine learning [18]. Other approaches taking advantage of quantitative structure-property relationships (QSPR) [19–21], the multiple linear regression (MLR) [19,22], partial least square regression (PLS) [23], support vector machines (SVMs) [22] and neural networks (NNs) [20] have been reported for predicting the solubility of C60 fullerenes in different organic solvents. Although these models offer quite an acceptable estimate suitable for screening of new solvents, they all rely on the sets of molecular descriptors characterizing solute-solvents properties. Very often these descriptors have no simple physical meaning and the predictive models can seriously suffer from the over-fitting problem in the training phase.

The different philosophy relies on the calculating of bulk phase equilibria from the first principles using well-established quantum chemistry approaches. Among these methods COSMO-RS (Conductor like Screening Model for Real Solvents) [24–26] offers an attractive alternative to linear and non-linear chemometrics approaches. Although it was originally formulated for computing thermodynamic properties of bulk liquid systems [27], it was extended also for treatment of interphases [28,29]. Particularly solids solubility [30] in organic solvents and their mixtures was the subject of significant interests [31–37]. Although the obtained results are of relatively poor or at most medium accuracy, they still offer valuable insight into the solid-liquid equilibrium phenomena. It is however important to emphasize the uniqueness of solids solubility modelling as requiring a proper thermodynamic formulation. Some basic, but important remarks are provided in the forthcoming paragraph.

Thermodynamic conditions for equilibrated mixtures of a solid–liquid saturated solution are characterized by equal values of chemical potential of pure solute solid phase, μ_i^{solid}, and of the solute in the saturated solution, μ_i^l. Formally this can be expressed as follows:

$$\mu_i^l = \mu_i^{solid} \tag{1}$$

$$\mu_i^{0,l} + RTln\left(\gamma_i^l\left(x_i^l\right)\cdot x_i^l\right) = \mu_i^{0,solid} + RTln\left(a_i^{solid}\right), \tag{2}$$

where $\mu_i^{0,l}$, $\mu_i^{0,solid}$ are the reference state chemical potentials, $\gamma_i^l\left(x_i^l\right)$ is an activity coefficient at the given value of mole fraction of solute in saturated solution, x_i^l, R is the universal gas constant and T denotes temperature. The thermodynamic reference state for the solute in the solution is often defined as the pure compound in a hypothetical super-cooled melt state at given temperature. On the other hand, in chemical practice often the activity of the solid phase is set to unity, which means that the reference state for the solid phase is the solid phase itself. Then, the solute activity in the saturated solution represents simply the difference between values of chemical potentials of chosen thermodynamic reference states, which means that

$$ln\left(a_i^l\left(x_i^l\right)\right) = ln\left(\gamma_i^l\left(x_i^l\right)\cdot x_i^l\right) = \frac{1}{RT}\left(\mu_i^{0,solid} - \mu_i^{0,l}\right). \tag{3}$$

Hence, activity of the solid phase, which obviously is different from unity, is exclusively solute related and in any saturated solution has exactly the same value provided that the same solid form is preserved. This of course does not mean that solubility represented by the mole fraction of a given solute is the same in different solvents because the activity coefficients are local properties related to concentration and solvent type, $a_i^l\left(x_i^l\right)$. Of course, experimental determination of solubility allows for quantification of activities if activity coefficients are known. Particularly, in the case of ideal solutions solubility equals the activity of the solute and can be deduced just from calorimetric measurements. In the case of real solvents varying solubility simply stands for deviations from the Raoult law and such non-ideality is accounted by the values of activity coefficients. Due to the fact that chemical potential by definition is a molar value of Gibbs free energy at a given temperature and pressure, Equation (3) can be rewritten in terms of pure solute properties leading to the following formula:

$$ln\left(\gamma_i^l\cdot x_i^l\right) = -\frac{\Delta G_i^{fus}(T)}{RT} = \frac{1}{RT}\left(T\Delta S_i^{fus}(T) - \Delta H_i^{fus}(T)\right) \tag{4}$$

where ΔG_i^{fus} is the value of a hypothetical partial molar Gibbs free energy of the melt at the system temperature and pressure. This value is temperature dependent and becomes zero for the pure solute at its melting point. Knowledge of the temperature related change of heat capacities of solid and melt solute $\Delta C_{p_i}^{fus}(T) = C_{p_i}^{l}(T) - C_{p_i}^{solid}(T)$, allows for computing entropic and enthalpic contributions to the values of fusion Gibbs free energy by the following basic definitions:

$$\Delta S_i^{fus}(T) = \Delta S_i^{fus}(T_m) + \int_{T_m}^{T} \frac{\Delta C_{p_i}^{fus}}{T} dT \tag{5}$$

$$\Delta H_i^{fus}(T) = \Delta H_i^{fus}(T_m) + \int_{T_m}^{T} \Delta C_{p_i}^{fus} dT \tag{6}$$

where T_m represents melting temperature of pure solute. Combining Equations (4) and (5) leads to the following thermodynamically rigorous relationship (if no phase transition occurs between different solid states):

$$ln\left(a_i^l\right) = -\frac{\Delta H_i^{fus}(T_m)}{RT}\left(\frac{T}{T_m} - 1\right) - \frac{1}{RT}\int_{T_m}^{T} \Delta C_{p_i}^{fus} dT + \frac{1}{R}\int_{T_m}^{T} \frac{\Delta C_{p_i}^{fus}}{T} dT \tag{7}$$

Unfortunately, direct application of this equation is impossible due to the lack of experimental data. With contemporary thermo-analytical techniques the temperature related values of isobaric heat capacity of the solid can be determined below the melting temperature. Additionally $C_{p_i}^{l}(T)$, characterizing the solute melt state, can be measured above the melting temperature, if the compound does not decompose. Unfortunately, the super-cooled melt system used as a reference state is in practice experimentally inaccessible far away from the melting temperature, as it is the case in the instance of typical solubility measurements. Hence, to overcome this difficulty several simplifications of Equation (7) were proposed enabling for extrapolation of heat capacities. First of all, it is possible to assume that $\Delta C_{p_i}^{fus}(T)$ is small in comparison to the other terms and is omitted by imposing a zero value. This means that the enthalpy of fusion is independent of the temperature. This seems to be reasonable for temperatures not too remote from the melting point [38,39]. On the other hand, there are strong suggestions that in general this represents serious oversimplification [40–42] since change of the heat capacity upon phase transition is very often significant. Hence, the alternative proposed by Hildebrand and Scott [43,44] assumes that $\Delta C_{p_i}^{fus}(T)$ is constant and can be approximated by the entropy of fusion at the melting temperature. In turn, this value can be approximated by the ratio of fusion enthalpy and system temperature. Although this assumption is not exceptionally accurate, it has been shown [41] that the true value of $\Delta C_{p_i}^{fus}(T)$ is generally much closer to $\Delta S_i^{fus}(T)$ than to zero. The importance of this assumption was validated by Moller [45] reporting a mean difference of 10% between these two simplifications. However, a more systematic study on complex substances did not confirm this notion since no significant influence has been found on the overall solubility predictions [46] in relation to assumptions of $\Delta C_{p_i}^{fus}(T)$ model. There are also different semi-empirical models developed for characteristics of solid-liquid heat capacities enabling much higher precision of solute activity estimation and thermodynamically rigorous extrapolation over a broad range of temperatures [47,48].

On the other hand, Equation (7) plays an important role in the area of theoretical solubility predictions and since it defines the problem with two unknowns coming from different sources all the above comments are valid also in this context. Many theoretical models were developed for estimating activity coefficients based on which solubility can be deduced. Among many theoretical approaches the COSMO-RS theory offers a very attractive way of chemical potential characteristics in bulk systems. Foundations of this theory are available in original papers [24–27], so details will be omitted here. It is suffice to remind that the values of the chemical potentials of compounds in the solvents are calculated by an iterative solution of the following equation:

$$\mu_i^l(\sigma) = -\frac{RT}{a_{eff}}ln\left[\int P_i(\sigma\prime)\cdot exp\left(\frac{a_{eff}}{RT}\left(\mu_i^l(\sigma\prime) - E(\sigma, \sigma\prime)\right)\right)d\sigma\prime\right], \tag{8}$$

where σ represents the charge density of molecular segment, $E(\sigma,\sigma')$ is the sum of electrostatic and hydrogen bonding contributions to the total energy of the system computed as pair-wise additive interactions of molecular surface segments of charge density σ. The molecular surface is defined by the molecule polarization after immersing in dielectric continuum or conductor. This solvation model [25] results in a so called σ-profile, $P_i(\sigma)$, representing simply a histogram of surfaces of a given charge density. The most important information of the molecular polarity and intermolecular interactions is directly encoded in σ-profile. The electrostatics (ES), hydrogen bonding (HB) and van der Waals interactions of contacting surface pieces of charge density σ and σ' are computed based on the pair-wise additivity assumption. The detailed definitions are of less importance here, except for the fact that they comprise scaling factors and adjustable model parameters fitted to a large number of thermodynamic data [25,26]. After integrating the segments chemical potentials over the surface of the i-th compound in the multicomponent system, the value of total chemical potential is computed by inclusion of a combinatorial term:

$$\mu_i^l = \int P_i(\sigma) \cdot \mu_i^l(\sigma) d\sigma + \mu_{COMB,i}^l(\sigma). \tag{9}$$

The first component plays the role of the residual part and the latter accounts for size and shape effects of solute and solvent [24,27]. Since the above equation offers a general way of computing chemical potential of any system, it is directly applicable to liquids solubility prediction. However, for solid solutes a generalization was proposed by accounting also the fusion contribution:

$$RTlog\left(x_i^{l,(iter+1)}\right) = \left[\mu_i^{o,l} - \mu_i^l\left(x_i^{l,(iter)}\right) - \max\left(0, \Delta G_i^{fus}(T)\right)\right]. \tag{10}$$

This solubility equation is solved iteratively until self-consistency criterion is met by re-computing values of the chemical potentials dependent on the mole fraction of solute in solution at the current iteration. This leads to systematic improvement of the initial guess of solubility representing solubility at infinite dilution in given solvent or solvents mixture.

The solubility in Equation (10) can be used also in the reversed manner for Gibbs free energy of fusion computations. Based on COMSO-RS derived chemical potential and experimental solubility fusion data are directly available. This value is to be treated as apparent Gibbs free energy of fusion, which knowledge, along with computed values of chemical potential, is sufficient for exact reproduction of the experimental solubility.

The main advantage of using COSMO calculations for calculating phase equilibria is that the results are obtained from the first principles. However, application of this approach for solid solutes solubility prediction, apart from its own shortcomings of estimation of chemical potentials, is also prone to fusion related inaccuracies as discussed above. The goal of this paper is three-fold and inherently associated with Equation (10) First of all, the critical evaluation of quality of predicted values of C60 solubility in 180 net solvents is undertaken. Then, the origin of the observed inaccuracies is discussed in the light of the values of Gibbs free energy of fusion derived from solubility data. Decomposing these data into meaningful contributions provides insight into the source of observed discrepancies between predicted and computed data. Finally, classification of solvents using the consonance solvents approach is proposed and solubility data predicted based on this idea are confronted against experimental data.

2. Materials and Methods

2.1. Solubility Dataset

Solubility data of C60 in a variety of solvents were compiled by different authors for modeling purposes [18,19,21,49,50]. In few cases, some discrepancies were found and averaged values were used in this paper. In Tables 1–3 all experimental values are provided.

Table 1. The list of type A of consonance reference solvents suggested according to modest and fine criterions of solvents classification. APE (absolute percentage error) quantifies of solubility computations if the given solvent is used as a reference solvent within the consonance group. Best solvent within each group is marked in bold face. Fluidization term (FLUID) is given in kcal/mol.

Solvent name	$\log(x^{exp})$	modest	APE	fine	APE	FLUID
Pentane	−5.10	A	19.7%	A1	5.8%	4.02
Methanol	−4.29		18.0%		4.8%	3.90
2-Chloropropane	−5.40		13.3%		2.5%	3.55
1-Chloropropane	−4.61		12.1%		2.3%	3.46
1,1,2-trichlorotrifluoroethane	−3.69		11.3%		2.3%	3.39
1-Chloro-2-methylpropane	−4.30		10.7%		2.6%	3.33
2-Chloro-2-methylpropane	−4.40		10.5%		2.7%	3.32
Cyclohexane	−3.80		9.0%	A2	5.0%	3.18
1-Bromopropane	−4.20		8.7%		4.7%	3.15
Tetrahydrothiophene	−4.38		8.7%		4.6%	3.14
2-Methylpentane	−5.35		8.6%		4.6%	3.14
2-Bromopropane	−4.79		7.8%		3.8%	3.05
3-Methylpentane	−4.95		7.4%		3.4%	2.99
Bromoethane	−4.50		6.5%		2.6%	2.87
Chloroform	−4.21		6.5%		2.6%	2.88
Silicon chloride	−3.89		6.4%		2.5%	2.85
Butan-2-ol	−5.57		5.7%		2.3%	2.73
1,1,1-Trichloroethane	−5.57		5.6%		2.3%	2.72
1,2-Dichlorodifluoroethane	−5.31		5.6%		2.3%	2.71
Ethanol	−5.91		5.6%		2.3%	2.69
2-Methylheptane	−3.80		5.5%		2.3%	2.68
Octane	−5.20		5.5%		2.3%	2.68
Isooctane	−4.71		5.4%		2.5%	2.65
1-Bromo-2-methylpropane	−5.70		5.4%		2.5%	2.64
Tetrahydrofuran	−2.80		5.4%		2.7%	2.62
Acetone	-5.40		5.4%		2.8%	2.61
Hexane	−4.98		5.4%		2.8%	2.61
Propan-2-ol	−6.34		5.4%		3.1%	2.58
Diiodomethane	−6.70		5.4%		3.0%	2.59
1-Iodopropane	−5.90		5.5%	A3	5.2%	2.54
1-Propanol	−5.12		5.5%		4.8%	2.51
2-Iodopropane	-4.89		5.5%		4.7%	2.51
Carbon tetrachloride	−4.48		5.6%		4.6%	2.50
2-Bromo-2-methylpropane	−3.70		5.7%		4.2%	2.46
n-Heptane	-4.82		5.8%		4.1%	2.45
1-Butanol	−5.32		6.7%		3.0%	2.27
Iodoethane	−3.10		6.7%		3.0%	2.27
Nonane	−4.78		6.7%		3.1%	2.28
1-Chloro-2-methylpropene	−3.66		6.5%		3.2%	2.31
cis-1,2-Dimethylcyclohexane	−4.30		7.9%		2.9%	2.11
Iodomethane	−4.19		8.4%		3.0%	2.06
trans-1,2-Dimethylcyclohexane	−4.64		8.5%		3.0%	2.05
1,2-Dimethylcyclohexane	−4.64		8.5%		3.0%	2.05
1-Iodo-2-methylpropane	−5.00		8.6%		3.1%	2.04
Pentan-2-ol	−5.36		9.0%		3.4%	2.00
Dibromomethane	−3.19		9.1%		3.5%	1.99
Decane	−3.90		9.3%		3.7%	1.97
Methylcyclohexane	−4.60		9.4%		3.7%	1.97
Thiophene	−3.01		9.5%		3.8%	1.96

Table 2. The list of type B of consonance reference solvents suggested according to modest and fine criterions of solvents classification. Notation is the same as in Table 1.

Solvent name	$\log(x^{exp})$	modest	APE	fine	APE	FLUID
Pentan-3-ol	−7.04	B	15.1%	B1	3.0%	1.84
Silicon bromide	−5.50		15.0%		2.9%	1.83
2-Iodo-2-methylpropane	−3.78		14.9%		2.8%	1.83
1-Octanol	−6.42		14.1%		2.3%	1.78
1-Pentanol	−7.00		13.4%		2.0%	1.74
Trichloroethylene	−3.39		13.2%		1.8%	1.72
Dichloromethane	−4.01		13.0%		1.8%	1.71
Cyclopentyl bromide	−4.08		12.9%		1.8%	1.70
2-methylphenol	−6.52		12.6%		1.7%	1.68
1-Hexanol	−5.00		12.5%		1.7%	1.67
Ethylcyclohexane	−4.78		12.4%		1.8%	1.67
Tetrachloroethylene	−4.50		12.2%		1.8%	1.65
Cyclohexyl chloride	−3.41		12.1%		1.9%	1.64
1,2-Dichloropropane	−4.81		10.4%		3.2%	1.49
Bromochloromethane	−5.18		9.8%		3.7%	1.44
Tetradecane	−5.28		9.6%		4.0%	1.42
1,1,2-Trichloroethane	−3.74		9.2%		4.7%	1.37
Carbon disulfide	−2.14		9.2%		4.7%	1.37
Benzene	−3.14		8.9%	B2	5.2%	1.34
1,2-dibromoethene	−5.38		8.8%		5.4%	1.33
1,2-Dibromopropane	−3.40		8.4%		2.8%	1.26
Cyclohexene	−4.50		8.3%		2.8%	1.26
1,2-Dibromoethylene	−3.79		8.3%		2.7%	1.24
Bromobutane	−3.30		8.2%		2.6%	1.24
nitroethane	−3.30		7.9%		2.5%	1.17
1-Methyl-1-cyclohexene	−4.60		7.9%		2.5%	1.16
Octanoic acid	−4.41		7.8%		2.5%	1.14
1,3-Dichloropropane	−4.30		7.8%		2.5%	1.13
Fluorobenzene	−3.22		7.8%		2.6%	1.11
1,2-Dichloroethane	−4.69		7.7%		2.8%	1.08
1,2-Dibromoethane	−5.18		7.5%		4.0%	0.93
Iodobenzene	−3.20		7.5%		4.0%	0.93
Nonan-1-ol	−4.15		7.5%		4.2%	0.92
Butanoic acid	−5.05		7.6%		5.4%	0.83
Decan-1-ol	−3.99		7.6%		5.9%	0.80
tert-Butylbenzene	−4.03		7.6%		4.4%	0.80
1,3-Dichlorobenzene	−2.60		7.6%	B3	4.3%	0.79
Cyclohexyl bromide	−2.80		7.7%		3.6%	0.75
Bromoform	−4.19		7.7%		3.6%	0.75
1,3-Dibromopropane	−4.90		7.9%		3.2%	0.72
Bromoheptane	−3.09		8.0%		2.8%	0.68
Undecan-1-ol	−6.65		8.3%		2.4%	0.64
trans-Decahydronaphthalene	−4.20		8.3%		2.4%	0.64
iso-Propylbenzene	−3.40		8.4%		2.2%	0.63
1,2-Dimethylbenzene	−3.25		8.6%		2.1%	0.60
1,3,5-Trimethylbenzene	−2.38		8.6%		2.1%	0.59
Bromobenzene	−2.30		8.7%		2.1%	0.58
sec-Butylbenzene	−2.76		8.7%		2.1%	0.58
N-propylbenzene	−3.62		8.9%		2.1%	0.57
Chlorobenzene	−2.81		8.9%		2.1%	0.57
Nonanoic acid	−5.72		9.0%		2.1%	0.56
1,5-Pentanediol	−3.40		9.5%		2.4%	0.52
Propargyl bromide	−4.82		9.6%		2.5%	0.51
Propylene glycol	−12.49		9.8%		2.6%	0.49
Toluene	−3.29		10.0%		2.7%	0.48
cis-Decahydronaphthalene	−3.50		10.7%		3.3%	0.43
Ethylbenzene	−3.61		10.8%		3.3%	0.43
Bromooctane	−2.59		11.3%		3.8%	0.39
n-Butylbenzene	−3.70		11.8%		4.3%	0.36
1,4-Dimethylbenzene	−2.91		12.1%		4.6%	0.35

Table 3. The list of type B of consonance reference solvent suggested according to modest and fine criterions of solvents classification. Notation is the same as in Table 1.

Solvent name	$\log(x^{exp})$	modest	APE	Fine	APE	FLUID
1,3-Dimethylbenzene	−3.09	C1	4.9%	C	12.4%	0.33
Dodecane	−3.31		4.5%		20.3%	0.32
Pentanoic acid	−4.50		4.2%		20.0%	0.31
1-Bromo-3-chloro-benzene	−2.78		4.2%		20.0%	0.31
Acrylonitrile	−5.20		3.5%		18.7%	0.24
2-Methoxyethyl ether	−5.00		3.5%		18.7%	0.24
Trichlorotoluene	−2.20		3.7%		19.1%	0.26
1,4-Butanediol	−6.19		2.4%		15.9%	0.09
1,3-Diiodopropane	−2.90		2.4%		15.8%	0.08
Benzaldehyde	−3.40		2.4%		15.7%	0.08
Cyclohexyl iodide	−2.60		2.3%		15.1%	0.04
1,2,4-Trichlorobenzene	−3.21		2.3%		15.0%	0.03
2-Methylthiophene	−3.91		2.3%		14.8%	0.02
Propionic acid	−5.74		2.5%		14.5%	0.00
Phenyl isocyanate	−3.40		2.5%		14.3%	−0.01
Hexanoic acid	−4.26		2.6%		14.2%	−0.02
Heptanoic acid	−4.98		2.7%		14.1%	−0.04
Styrene	−3.92		2.7%		14.1%	−0.04
N,N-Dimethylformamide	−5.18		2.8%		14.0%	−0.04
2,3,4,5-tetrahydro-1H-azepine	−2.70		3.1%		13.8%	−0.06
1,2,3,4-Tetramethylbenzene	−3.42		3.6%		13.5%	−0.09
1,2,3-Trichloropropane	−5.40		3.9%		13.4%	−0.11
1,3-Propanediol	−6.60		4.1%		11.3%	−0.35
Benzyl chloride	−2.01		3.8%		11.2%	−0.37
1-Bromotetradecane	−2.53		3.8%		11.2%	−0.37
N,N-Dimethylaniline	−2.70		3.7%		11.2%	−0.38
Acetonitrile	−5.16		2.7%		10.7%	−0.46
Pyridine	−2.91		2.5%		10.6%	−0.49
1-Bromooctadecane	−8.78		2.3%		10.5%	−0.51
Thiophenol	−3.10		2.2%		10.4%	−0.54
1-Bromo-2-chloro-benzene	−3.00		2.2%		10.4%	−0.54
1,5,9-Cyclododecatriene(Z,Z,E)	−4.27		2.2%		10.3%	−0.57
1,5,9-Cyclododecatriene(E,E,Z)	−7.03		2.2%		10.3%	−0.58
1,2-Dichlorobenzene	−2.60		2.3%		10.3%	−0.58
Benzyl bromide	−3.00		2.5%		10.3%	−0.60
1,2-Dibromocyclohexane	−3.79		2.7%		10.4%	−0.61
Anisole	−4.21		2.9%		10.4%	−0.62
Nitrobenzene	−4.22		4.1%		10.7%	−0.70
1,2,4-Trimethylbenzene	−3.13	C2	3.7%		10.8%	−0.72
1,4-dioxane	−5.62		2.8%		11.1%	−0.78
Acetic acid	−5.79		2.1%		11.6%	−0.86
1,2,3,5-Tetramethylbenzene	−2.67		2.2%		11.4%	−0.83
Benzonitrile	−3.25		2.0%		12.2%	−0.94
1,2,3-Tribromopropane	−3.98		2.0%		12.2%	−0.93
m-Bromoanisole	−2.11		2.3%		12.4%	−0.97
2-Nitrotoluene	−3.41		3.5%		13.2%	−1.04
p-Bromoanisole	−2.55	C3	3.3%		13.5%	−1.07
1,3-Diphenylacetone	−2.54		2.9%		13.7%	−1.09
1-Bromo-2-methylnapthalene	−6.40		2.9%		13.7%	−1.09
3-Nitrotoluene	−3.00		2.9%		13.8%	−1.09
N-methyl-2-pyrrolidone	−4.00		2.6%		14.3%	−1.13
2,4,6-Trimethylpyridine	−6.27		2.4%		15.7%	−1.22
1-Methylnaphthalene	−2.10		2.4%		15.7%	−1.23
n-Butylamine	−4.60		2.8%		16.5%	−1.27
1-Chloronaphthalene	−2.10		3.1%		17.0%	−1.30

Table 3. *Cont.*

Solvent name	$\log(x^{exp})$	modest	APE	Fine	APE	FLUID
1,2-Dimethylnaphthalene	−7.54	C4	3.2%		17.1%	−1.30
2,6-Dimethylnaphthalene	−1.90		1.9%		18.4%	−1.37
Quinoline	−3.89		1.7%		19.8%	−1.43
Aniline	−3.79		3.9%		23.2%	−1.58
1-Phenylnaphthalene	−7.10		9.1%	C′	9.1%	−1.75
piperidine	−2.27		6.0%		6.0%	−2.06
pyrrolidine	−5.20		8.1%		8.1%	−2.21
N-methylaniline	−3.19	D1	6.0%	D	6.0%	−0.44
1,1,2,2-Tetrachloroethane	−5.02		2.3%		2.3%	−0.23
Tetralin	−3.51		2.3%		2.3%	−0.23
1,2-Dibromobenzene	−3.40		2.4%		2.4%	−0.25
1,3-Dibromobenzene	−2.40		2.5%		2.5%	−0.20
1,2,3-Trimethylbenzene	−2.24		8.2%		8.2%	0.13
Cyclopentane	−4.77					
Nitromethane	−3.18					
Water	−12.49					

2.2. Thermodynamic Data of C60 Fusion

So far the fusion of fullerene C60 has not been observed at any temperature and existing estimate values do not offer consensus. For example, the following data were reported: $T_m > 950$ K [51], $T_m \approx 1200$ K [52] or $T_m \approx 2023$ K [53]. These temperatures are rather debatable since C60 sublimates at temperature between 700 and 945 K [54], depending on the experimental protocol used for the measurements. However, the enthalpy of fusion can be obtained from the available experimental enthalpy of sublimation and an estimated enthalpy of vaporization of the liquid fullerene. This leads to different estimates depending on the assumed values [49]. Here, all solubility computations rely on the values of fusion provided by Kulkarni et al. [55]. Hence, 25.77 kJ/mol was used for approximate fusion enthalpy as calculated from crystal energy in toluene assessed using solubility parameters approach [55]. Instead of melting temperature the sublimation one is assumed here, $T_m = T_{sub} = 750$ K [56]. These values were successfully used for C60 solubility prediction in solvent mixtures with an aid of Wohl's equation and Scatchard–Hildebrand theory [57].

2.3. COSMO-RS Computations

The structures of solute and solvent molecules were represented by sets of relevant conformations generated using COSMOconf 4.2 and Turbomole 7.0 as the default engine for geometries optimization. The obtained conformers used further for characteristics of bulk systems had their geometries fully optimized using BP functional and TZVP basis set. All structures were generated both in the gas phase and including environment effects via the COSMO-RS [25] solvation model. For solubility computations TZVPD-FINE level was used, which corresponds to single point calculation with TZVPD basis set and the same density functional based on previously generated geometries. The BP-TZVPD-FINE_19.01 parameterization was used for all physicochemical properties computations as implemented in COSMOtherm [Version 19.0.1 (Revision 5259)].

3. Results and Discussion

The starting point of this project was the frustrating observation of extremely high deviations of values computed using the COSMO-RS approach compared to experimental solubility data of fullerene C60 dissolved in 180 net solvents. This is documented in Figure 1 by the distribution marked with black crosses representing results of default computations done in COSMOtherm on fine level. Practically, there is no correlation between predicted and measured values ($R^2 = 0.364$). Although the mean absolute percentage error, MAPE = 23% seems to be acceptable, only half of the systems is characterized with absolute percentage error, APE, less than 20%. In the case of cyclopentane the percentage error, PE, exceeds 55% and for pyrrolidine reaches −73%. These values characterize deviations expressed in the decadal logarithm. This means in practice one order of magnitude under- or overestimation of mole fractions of C60 in these solvents. For the majority of solvents the solubility of C60 is significantly overestimated by COSMO-RS without any detectable trend related to molecular structure of the solvents. For some alcohols, as for example 1,4-Butanediol, computed C60 solubility perfectly match experimental ones, APE = 1.0%. For other solvents of this type, as for example methanol, a significant discrepancy with APE = 32.2% is observed. Generally poorer performance is noticed for less polar systems containing aliphatic chains and better for systems being able to form hydrogen bonds and with high contributions of electrostatic interactions. For detailed inspection of the origin of this situation the reversed solubility computations were performed for all systems and apparent Gibbs free energy of fusion were generated.

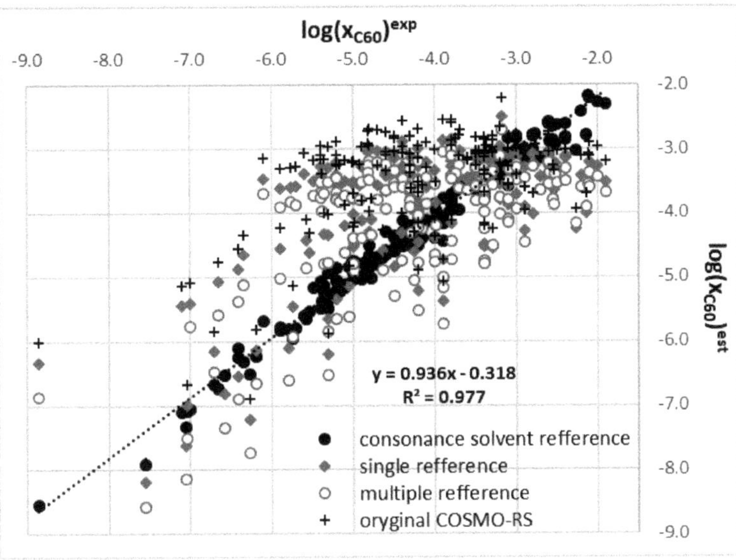

Figure 1. The correspondence between computed and experimental values of fullerene C60 solubility in 180 net organic solvents. Statistics of regression curve denotes accuracy of consonance solvent selection according to fine criterion.

3.1. Computations of C60 Fusion from Solubility Measurements

As it was mentioned in the introduction that the procedure of solubility computation in COSMO-RS can be reversed for estimating adequate values of fusion Gibbs free energy based Equation (10) allowing for straightforward determination of ΔG_{fus} values. This means that if a proper value of such fusion affinity is available normal solubility computations using COSMOtherm would reproduce exactly experimental data. This option relies simply on the reference solubility approach in which chemical potential of the solute is determined based on experimental concentration of saturated solution and computed values of the equilibrium activity of the solute $a_i^l(x_i^{sat})$. The computed values characterizing Buckminster fusion for the set of available 180 solubility values are presented in Figure 2. Presented distribution of the apparent Gibbs free energy of fusion univocally documents that this value is very strongly dependent on the nature of the solvent. For generalization of the discussion, instead of absolute values of solubility-derived fusion, the relative value with respect to calorimetrically measured data will be used and termed here as fluidization contribution, $FLUID = \Delta G_{C60,sat}^{fus,COSMO-RS} - \Delta G_{C60}^{fus,cal}$, where $\Delta G_{C60}^{fus,cal}$ equals 3.71 kcal/mol at room temperature [55]. The obtained result presented in Figure 2 is rather unexpected since in an ideal situation FLUID should be at least constant if not equal to zero for any solvent at constant temperature and pressure. Evidently this is not the case for fullerene C60 since the deviations from calorimetric fusion value range from –3.4 kcal/mol for nitromethane up to +4.9 kcal/mol in the case of cyclopentane. Data for other few extreme cases are also provided in Figure 2. Such pronounced variability of apparent ΔG_{fus} cannot be addressed solely to improper representation of the heat capacities. Additionally, problems with measurements of melting temperature and heat of fusion of C60 cannot justify such strong irregularities.

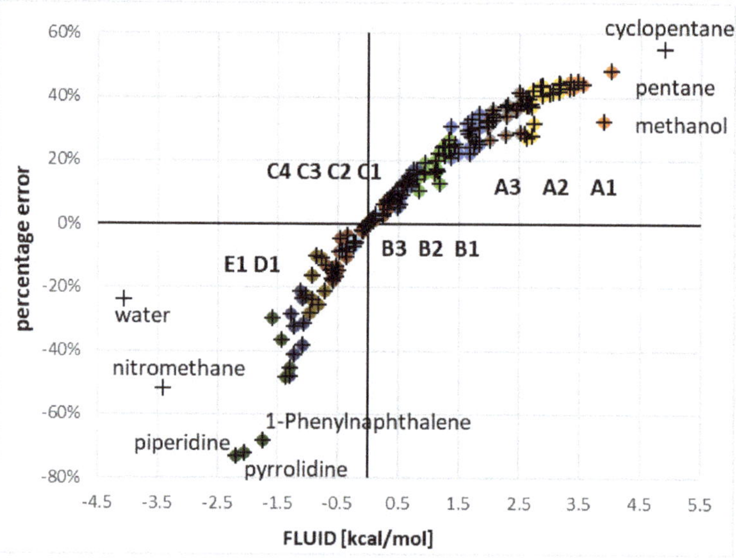

Figure 2. Correspondence between percentage error and values of fluidization term of Buckminster fusion derived based on experimental solubility data and calorimetric measurements (FLUID = $\Delta G_{C60,sat}^{fus,COSMO-RS} - \Delta G_{C60}^{fus,cal}$). Data points are colored according to solvents consonance classified using fine criterion. Codes represent subclasses enlisted in Tables I-III.

According to equation 9 the relationship between $\log(x_{C60}^{sat})$ and ΔG_{fus} is linear and if inaccuracy of the fusion value equals 1 kcal/mol than the error of solubility expressed in units of decadal logarithm of mole fraction at room temperature will be equal to $RT\ln(1) \approx 0.74$. This means that for proper estimation of solubility a very high accuracy of ΔG_{fus} is required exceeding the so called chemical accuracy, which is of the order of 1 kcal/mol or 0.03 eV [58]. However, in the case of C60 the solvent imposed alterations are much higher, variable and not negligible. For 103 solvents FLUID exceeds 0.5 kcal/mol and for 76 is lower than –0.5 kcal/mol suggesting that C60 is to be considered as a rather non-ideal agent in the majority of solvents despite low concentrations at saturated conditions. In fact, this is supposed to be the reason of very poor C60 solubility predictions by COSMO-RS, as documented in Figure 1. This means that activities coming from the COSMO-RS approach are inaccurately computed not accounting for all necessary contributions characterizing saturated solutions. The span of solubility values is very broad encompassing solvents in which C60 is practically insoluble as for example water ($\log(x_{C60}^{sat}) = -12.5$) and modestly miscible as it is in the case of 1-phenylnaphthalene $\log(x_{C60}^{sat}) = -1.9$.

It is not surprising that fluidization terms for cyclopentane and nitromethane have opposite signs due to differences in electronic structure affecting intermolecular interactions in bulk systems. Generally, in the case of non-polar structures as alkanes and cycloalkanes, the fluidization contribution is positive indicating higher affinity of solute to such solvents, what results in overestimation of solubility in the case of using to low ΔG_{fus} values. To the contrary, highly polar and rich in lone pairs systems, as for example nitromethane or heteroaromatic compounds, are characterized by negative values of FLUID term. Hence, the affinity of solute toward such solvents is lower than expected from uncorrected calorimetric values of G_{fus} and computed solubility will be underestimated. It is a bit surprising that methanol and also other alcohols were found in the same group as some alkanes or halogenated alkanes but this should be addressed probably to different sources of FLUID value and cancelation of such contributions as electrostatic, hydrogen boding and dispersion.

Before proceeding any further it is necessary to comment on the common phenomenon observed in the case of systems in which high affinity of solute is observed toward solvent molecules. In such cases, it is quite reasonable to expect that new solid forms can precipitate in the saturated solution in the form stable solvates. This in turn would give rise to a different activity for the solid phase due to altered thermodynamic properties of solid phase of different composition. Such multicomponent solids are well known in medicinal chemistry. It was already recognized that variations in the composition of the contents of the stomach/gut may affect the recrystallization kinetics between different solid-state forms of a drug [59,60]. This can be addressed not only to polymorphism but more often to solvates as for example it is the case for carbamazepine [61–63] and its cocrystals [60]. This aspect is also relevant to C60 solid–liquid equilibria in organic solvents. Unfortunately, in reports providing the values of solubility there are not sufficient information about solid phase present in the equilibrium with the saturated solutions. Although no direct clues about C60 solvates can be inferred from solubility measurements this aspect was not overlooked and attracted serious attention by many investigators [10–17,64–68]. For example, the dissolution properties of C60 in aromatic solvents were studied using the thermos-analytical approach [10,12] and many solid solvates were identified. This of course is relevant to the data provided in Figure 2 since documented dramatic influence of the thermodynamic properties of solid solvates might be addressed to this phenomenon of C60.

3.2. Reference Solvent Computations

The remedy on the lack of G_{fus} is supposed to be the so called reference solubility computation. It is simply an attempt of computing solubility in one solvent based on information of solubility in another. In practice, this is exactly the same procedure as the one used for computing the FLUID term but two solvents are declared in the input files. Since it is not known *a priori* which reference solvent is the most appropriate for given solvent all possible combinations were considered here. Hence, reference solubility was computed for all $180 \times 179 = 32{,}220$ solvents pairs. In each case one solvent was used as a reference one and solubility was computed for the other solvent. Based on obtained results MAPE was computed for each reference solvent. The best result of these computations was provided in Figure 1 as distribution marked with grey diamonds. It happened that the selection of either ethylbenzene, toluene, bromooctane, cis-decahydronaphthalene, propylene glycol, propargyl bromide, 1,5-Pentanediol or n-butylbenzene leads to similar quality of solubility prediction for the rest solvents. Unfortunately, this approach is unsatisfactory as evidenced by the plot collected by Figure 1, where ethylbenzene served as the single reference solvent system. Obviously, the diversity of solvents in the data set prevents from finding one optimal reference system. Hence, this option does not provide any sensible increase of overall accuracy of solubility predictions if a single set of reference solvents is applied to the whole data set. It is also worth mentioning that the reference solvent method cannot help in the cases if solid states in equilibrium with saturated solutions comprise solvates. Since, there is a vast gross of evidences [10–14,16,17,64–66,68] that Buckminster can interact with many solvents, it is quite clear that reference solubility cannot be used neither for serious improvement of predicted solubility data nor for treating cases of unknown fusion characteristics.

3.3. Consonance Solvents Classification

Fortunately, the situation is not as hopeless as it might seem from results presented so far since some interesting trends can be revealed after sorting data obtained based on the single reference solvent computations. The 180×180 matrix comprising APE values of every pair solvent-reference solvent was sorted systematically by rows and columns with criterion of ascending values of APE. After sorting rows, the columns were rearranged accordingly for keeping the same order of rows and columns. Finally, this resulted in a kind of heat map presented in Figure 3.

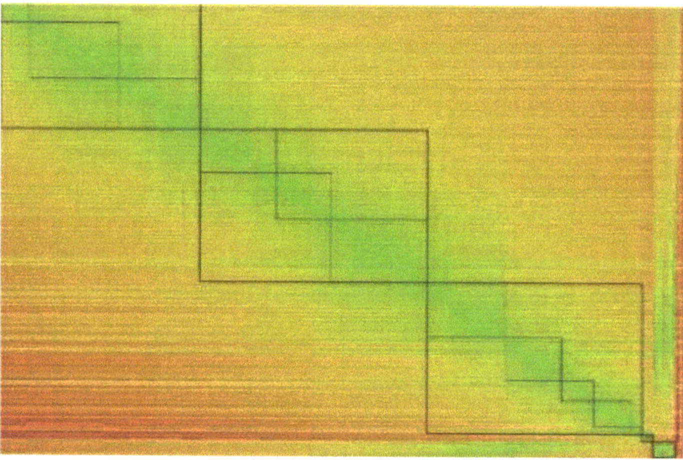

Figure 3. Graphical representation of consonance reference approach. Rows and columns ordered with increasing percentage error values were colored using green (low PE values) to red (high PE values) spectrum. Squares define consonance regions according to fine (PE < 10%) and modest (PE < 15%) criterions. The order of solvents is the same as provided in Tables 1–3.

This operation allows for grouping solvents according to their ability of playing the role of good reference solvent for other solvents found in the near neighborhood in Figure 3. Existence of such regions, encompassing solvents with high similarity in terms of FLUID values, is termed here as the consonance solvent approach. This means that on the map provided in Figure 3, for each case, it is possible to find suitable reference solvent providing quite accurate values of computed C60 solubility. Restricting the expected accuracy to the fine criterion for which APE < 10% of solubility prediction of every pair solvent-reference solvent results in rectangle regions marked with thinner black lines in Figure 3. Alternatively, reducing accuracy to APE < 15% leads to a much lower number of sections marked with bold black squares. In the former case, it is necessary to know thirteen solubility values for proper computation of solubility of the rest of the population. If the modest criterion is accepted only five measurements are indispensable for the same purpose. It happens that solvents in the middle of each class or subclass are the most suited reference solvents. However, three cases are out of this classification for which it is hardly possible to find suitable reference solvent. These are water, cyclopentane and nitromethane. These solvents are treated by COMOS-RS on different manner compared to the rest of the population that the consonance solvent approach is unable to improve solubility predictions. Particularly, water cannot be used as reference solvent due to ultra-low solubility of C60. Consequently, reference solubility predicts complete miscibility in majority cases if water is used as a reference system. Additionally, none of the organic solvents seem to be an appropriate reference for water with one exception. Using nitromethane as a reference solvent leads to 3.8% error of estimated C60 solubility in water. This is probably an accidental artefact. Cyclopentane and nitromethane are those solvents, which were found to be characterized by extreme values of fluidization term, which already anticipated their uniqueness.

The effectiveness of proposed classification is demonstrated in Figure 1 by a series marked with black circles. The mean absolute error is as low as 4.3% and the regression coefficient reached $R^2 = 0.977$ indicating very high accuracy of computed values. Reducing accuracy to the modest level results in an increase of the overall error up to 8.4% with reduced linearity of the trend to $R^2 = 0.904$. The coefficient of the linear regression equation provided in Figure 1 suggests that predicted solubility values are still slightly overestimated compared to measurements. The detailed classification with identified consonant solvents is provided in Tables 1–3.

According to the modest criterion, the first class of consonance solvents comprises 49 solvents. Alkanes and halogenated alkanes were mainly included here. However, surprisingly, also some light alcohols as methanol and ethanol were found within this class. Hexane is suggested as the best reference solvent for this class offering accuracy as good as MAPE = 5.4%. There are however many other solvents, which can be treated as consonance ones with similar applicability. For example, such commonly used solvents as tetrahydrofuran, acetone or propan-2-ol are promising substituents for hexane in reference solubility computations of C60 solubility. As it was enumerated in Table 1, for further increase of solubility predictions this class can be subdivided into three subclasses. This increases the accuracy by a factor of two, reducing MAPE to 2.1%. Hence, very accurate predictions of C60 solubility in all solvents collected in Table 1 are possible provided that three experimental measured values are available one for each subclass A1, A2 or A3. Interestingly, class A is characterized by positive and highest values of fluidization term being in the range from 1.36 kcal/mol for thiophene and 4.02 kcal/mol for pentane.

Similarly, collection of further 60 solvents provided in Table 2 can also be classified according to the consonance rule. Although the best reference solvent for class B is nonan-1-ol but also such solvents as 1,2-dichloroethane, 1,2-dibromoethane, iodobenzene, butanoic acid or decan-1-ol offer similar accuracy. The average percentage error is slightly higher in comparison with class A and is equal to 7.5%. As it is presented in Figure 2 class B is characterized by smaller values of fluidization term being within the range of 0.35 kcal/mol for 1,4-dimethylbenzene and 1.84 kcal/mol for pentan-3-ol. Class B also can be subdivided into three subclasses for increase of accuracy of computed C60 solubility.

Finally, in Table 3 the classification of the remaining 68 solvents is provided. Here are the solvents characterized either by close to zero values of fluidization term or negative values. This class is more heterogeneous compared to the previous two and consists of a larger selection of 59 solvents augmented with two small sets grouping three (class C') or six (class D) solvents. It has been found that 1,5,9-cyclododecatriene is the best reference solvent for class C offering 10.3% accuracy expressed in term of MAPE value. Alternatively, other solvents offer similar accuracy as for example 1-bromooctadecane, thiophenol, 1-bromo-2-chloro-benzene, 1,5,9-cyclododecatriene, 1,2-dichlorobenzene, benzyl bromide or 1,2-dibromocyclohexane. Class C', which merges with class C4 in case of fine classification, comprises solvents for which selection of piperidine seems to be the best reference solvent. Finally, selection of tetralin is suggested as consonance solvent for the final class.

3.4. Multiple Reference Solvent Computations

Results presented above tempted for testing if any benefit will result from solubility computations based on information of solubility in several solvents. This option, referred to as the multiple reference solvent approach, was extensively tested here for C60 solubility. There were several conducted trials with a varying number of solvents starting from two and ending on the maximum possible number in the current version of COSMOtherm. Combinations of suggested consonance solvents were used and also randomly selected ones but no significant benefit was achieved as a result of all these attempts. The best results found during this stage were provided in Figure 1 as distribution marked with open grey circles. The conclusion is quite simple and univocal. No sets of multiple reference solvents were found for which the quality of prediction would be comparable to the single consonance solvent approach.

3.5. Decomposition of Fluidization Term

In the context of the mentioned variability of fluidization term, a closer inspection of the factors contributing to its values seems to be valuable. Fluidization encompasses all contributions, which are inherently associated with dissolution but not included in the value of chemical potential of solute in the saturated solution computed by COSMO-RS. This might be related to improper accounting for entropy by the combinatory term. In addition, the residual portion of chemical potential might be the source of non-negligible FLUID values due to the inaccurate characteristics of some solvent specific aspects of solubilization including mixing, solvation and solvent structure alterations imposed by solute cavitation in the presence of the solute. Indeed, the incoherence of the COSMO-RS theory in the current formulation was already noticed in the case of strongly interacting species which resulted in formulation of the COMSO-RS-DARE approach [69]. This extension of the original method was introduced for proper evaluation of activity coefficients for the binary mixtures of carboxylic acids and non-polar components. Recently the author demonstrated the effectiveness of this approach for ethenzamide solubility predictions [70].

The quantification of fluidization contributions to the values of fusion Gibbs free energy is indispensable for a proper prediction of solids. To the author's best knowledge, this aspect was not raised so far, mainly because of the fact that there is no other substance, which solubility was so extensively studied as fullerene C60. In this case, conducting measurements in 180 solvents with such high chemical diversity, covering all types of solvents enables in depth analysis of the discussed aspect. For further exploring of the feature of the fluidization term, it seems to be valuable splitting it into combinatory contribution and residual one. Additionally, the latter term can be further decomposed into three major energetic contributions used by the COSMO-RS approach for chemical potential computations.

$$\Delta G_{fluid}(T) = \Delta G_{fluid}^{COMB}(T) + \Delta G_{fluid}^{RES}(T) =$$
$$= \Delta G_{fluid}^{COMB}(T) + \Delta G_{fluid}^{MSF}(T) + \Delta G_{fluid}^{HB}(T) + \Delta G_{fluid}^{vdW}(T) \tag{11}$$

where ΔG_{fluid}^{MSF} is the misfit part of the fluidization term coming from electrostatics interactions, ΔG_{fluid}^{HB} stands for hydrogen bonding contribution and ΔG_{fluid}^{vdW} accounts for all non-specific interactions expressed as van der Waals interactions. All four contributions can be estimated by scaling capabilities of COMSOtherm and enforcing corresponding scaling factors to be equal to zero. The result of separations of these contributions is provided in Figure 4.

First of all, the provided series show a broad trend suggesting that higher solubility is generally associated with lower values of fluidization term and such contributions as vdW and RES. On the contrary, the stronger the electrostatics the lower the solubility of C60. The remaining two contributions, namely HB and COMBI have much smaller values. Moreover, the origin of the observed variation of FLUID comes from the vdW term. It is somewhat compensated with misfit values for those solvents with higher solubility and consequently COMS-RS performs slightly better in such situations. Contrary to this, low solubility systems pose a real challenge to COSMO-RS.

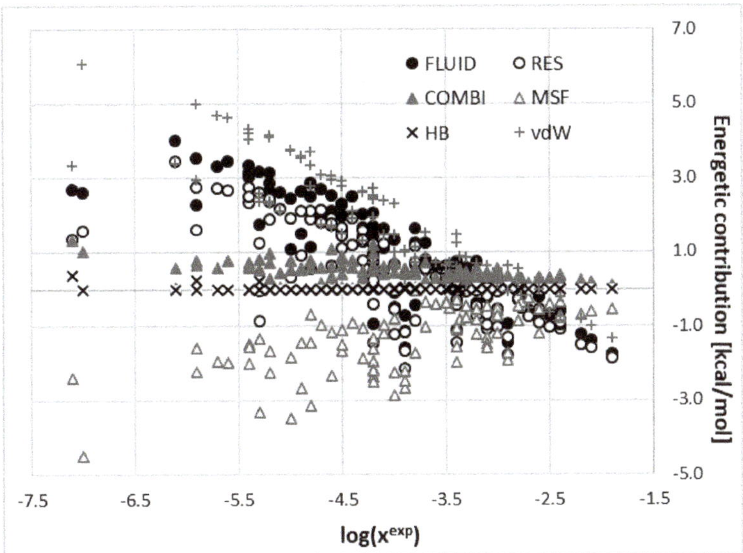

Figure 4. Relationship between different fluidization contributions and experimental solubility of C60 in studied solvents. MSF denotes the misfit part of fluidization term coming from electrostatics interactions (ΔG^{MSF}_{fluid}), HB stands for hydrogen bonding contributions (ΔG^{HB}_{fluid}), vdW accounts for all non-specific interactions expressed as van der Waals interactions (ΔG^{vdW}_{fluid} H), RES represents the residual part of the chemical potential by summing of these three terms, COMBI is the combinatorial part of chemical potentials and FLUID is the sum of all contributions.

Finally, these conclusions can be also supported by statistical analysis of the distributions by performing the principal component analysis (PCA) on covariance matrix. This step reveals that all contributions involved in Equation (11) can be reduced just to two statistically significant factors encompassing 98.5% of the total variance. Values of loadings provided in Figure 5 suggest that the whole fluidization term is mainly determined by its residual part and these two contributions are highly correlated with PC1 with correlation coefficient as high as –0.99 and –0.98, respectively. The major contribution to the residual part comes from dispersion and electrostatics influences RES in much lesser extent. It is really interesting to see vdW and MSF loading to PC2 not only because of the fact that they predominantly define this component, but also for their opposite signs of their contributions. This explains why so chemically diverse compounds were grouped into the same classes of consonance solvents. Indeed, increase of dispersion is compensated by electrostatic contribution to fluidization term and vice versa. Due to small values of both combinatorial term and hydrogen boding their overall importance is marginal for considered data set what was already documented in Figure 4.

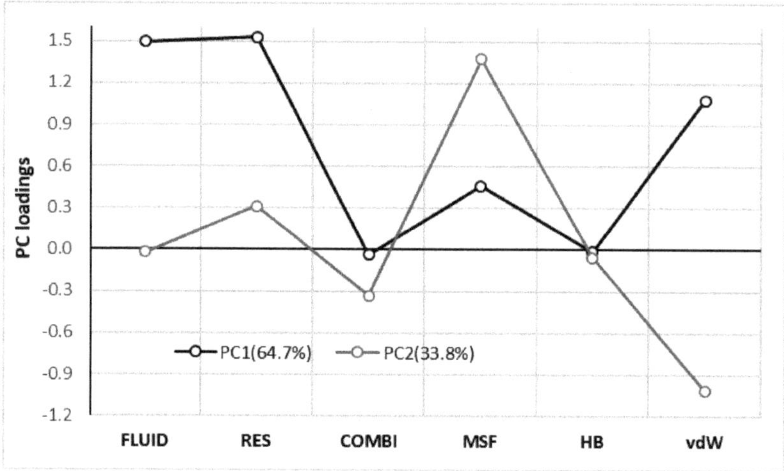

Figure 5. Results of principal component analysis of FLUID and its components including the residual part of the chemical potential (RES), combinatorial term (COMB), contributions coming from electrostatic (MSF), hydrogen boding (HB) and van der Waals (vdW) interactions.

4. Conclusions

Solubility of Buckminster dissolved in 180 net organic solvents was predicted using the COSMO-RS approach. This universal methodology, taking advantage of the first principle quantum chemistry computations augmented with statistical thermodynamics, allows in principle for characteristics of any bulk liquid systems or their interfaces. Formally, application to solid solubility requires data characterizing the fusion process. It was documented that from the perspective of the COSMO-RS approach it is indispensable to distinguish calorimetric contributions to Gibbs free energy of fusion from the fluidization term resulting from inappropriate characteristics of probably both residual and combinatorial parts of chemical potential. The higher the FLUID the lower the experimental solubility, which is also associated with a significant increase of error of computed solubility.

These negative conclusions came from performed computations stimulated for seeking some solution to this discouraging result. This goal has been achieved by proposal of classification of solvents into groups with similar values of fluidization term. It was possible to find sets of solvents used in the reference solvent computations protocol that lead to very accurate prediction of C60 solubility. However, it is not possible to define any set of reference solvents suitable for computation of the solubility for the whole set of solvents. As the rule of thumb solvents within interval of 1 kcal/mol of fluidization term can be mutually exchanged either as the solvent or the reference.

It is of course debatable if the observations made in this paper are of general nature or hold just for Buckminster dissolvent in net organic solvents. What is certain, however, is that observations were possible because of the richness of the experimental data available in the literature and further exploration is worth the effort.

Funding: This research received no external funding.

Acknowledgments: In this section you can acknowledge any support given which is not covered by the author contribution or funding sections. This may include administrative and technical support, or donations in kind (e.g., materials used for experiments).

Conflicts of Interest: The author declares no conflict of interest.

References

1. Rohlfing, E.A.; Cox, D.M.; Kaldor, A. Production and characterization of supersonic carbon cluster beams. *J. Chem. Phys.* **1984**, *81*, 3322–3330. [CrossRef]
2. Kroto, H.W.; Heath, J.R.; O'Brien, S.C.; Curl, R.F.; Smalley, R.E. C60: Buckminsterfullerene. *Nature* **1985**, *318*, 162–163. [CrossRef]
3. Vougioukalakis, G.C.; Roubelakis, M.M.; Orfanopoulos, M. Open-cage fullerenes: Towards the construction of nanosized molecular containers. *Chem. Soc. Rev.* **2010**, *39*, 817–844. [CrossRef] [PubMed]
4. Feng, L.; Liu, Z. *Biomedical Applications of Carbon Nanomaterials*; Biomedical Applications and Toxicology of Carbon Nanomaterials; John Wiley & Sons: Hoboken, NJ, USA, 2016; pp. 131–162, ISBN 9783527338719.
5. Bakry, R.; Vallant, R.M.; Najam-ul-Haq, M.; Rainer, M.; Szabo, Z.; Huck, C.W.; Bonn, G.K. Medicinal applications of fullerenes. *Int. J. Nanomed.* **2007**, *2*, 639–649.
6. Yadav, B.C.; Kumar, R. Structure, properties and applications of fullerenes. *Int. J. Nanotechnol. Appl.* **2008**, *2*, 15–24.
7. Beck, M.T.; Mándi, G. Solubility of C_{60}. *Fuller. Sci. Technol.* **1997**, *5*, 291–310. [CrossRef]
8. Nimibofa, A.; Newton, E.A.; Cyprain, A.Y.; Donbebe, W. Fullerenes: Synthesis and Applications. *J. Mater. Sci. Res.* **2018**, *7*, 22. [CrossRef]
9. Mojica, M.; Alonso, J.A.; Méndez, F. Synthesis of fullerenes. *J. Phys. Org. Chem.* **2013**, *26*, 526–539. [CrossRef]
10. Korobov, M.V.; Mirakian, A.L.; Avramenko, N.V.; Valeev, E.F.; Neretin, I.S.; Slovokhotov, Y.L.; Smith, A.L.; Olofsson, G.; Ruoff, R.S. C_{60} Bromobenzene Solvate: Crystallographic and Thermochemical Studies and Their Relationship to C_{60} Solubility in Bromobenzene. *J. Phys. Chem. B* **2002**, *102*, 3712–3717. [CrossRef]
11. Korobov, M.V.; Stukalin, E.B.; Mirakyan, A.L.; Neretin, I.S.; Slovokhotov, Y.L.; Dzyabchenko, A.V.; Ancharov, A.I.; Tolochko, B.P. New solid solvates of C_{60} and C_{70} fullerenes: The relationship between structures and lattice energies. *Carbon* **2003**, *41*, 2743–2755. [CrossRef]
12. Avramenko, N.V.; Mirakyan, A.V.; Neretin, I.S.; Slovokhotov, Y.L.; Korobov, M.V. Thermodynamic properties of the binary system C_{60}-1,3,5-trimethylbenzene. *Thermochim. Acta* **2000**, *344*, 23–28. [CrossRef]
13. Aksenov, V.L.; Tropin, T.V.; Kyzyma, O.A.; Avdeev, M.V.; Korobov, M.V.; Rosta, L. Formation of C_{60} fullerene clusters in nitrogen-containing solvents. *Phys. Solid State* **2010**, *52*, 1059–1062. [CrossRef]
14. Kyzyma, O.A.; Korobov, M.V.; Avdeev, M.V.; Garamus, V.M.; Petrenko, V.I.; Aksenov, V.L.; Bulavin, L.A. Solvatochromism and Fullerene Cluster Formation in C_{60}/N-methyl-2-pyrrolidone. *Fuller. Nanotub. Carbon Nanostruct.* **2010**, *18*, 458–461. [CrossRef]
15. Mikheev, I.V.; Khimich, E.S.; Rebrikova, A.T.; Volkov, D.S.; Proskurnin, M.A.; Korobov, M.V. Quasi-equilibrium distribution of pristine fullerenes C_{60} and C_{70} in a water–toluene system. *Carbon.* **2017**, *111*, 191–197. [CrossRef]
16. Snegir, S.V.; Tropin, T.V.; Kyzyma, O.A.; Kuzmenko, M.O.; Petrenko, V.I.; Garamus, V.M.; Korobov, M.V.; Avdeev, M.V.; Bulavin, L.A. On a specific state of C_{60} fullerene in N-methyl-2-pyrrolidone solution: Mass spectrometric study. *Appl. Surf. Sci.* **2019**, *481*, 1566–1572. [CrossRef]
17. Mikhail, V.K.; Andrej, L.M.; Avramenko, N.V.; Olofsson, G.; Smith, A.L.; Ruoff, R.S. Calorimetric Studies of Solvates of C_{60} and C_{70} with Aromatic Solvents. *J. Phys. Chem. B* **1999**, *103*, 1339–1346.
18. Gupta, S.; Basant, N. Predictive modeling: Solubility of C_{60} and C_{70} fullerenes in diverse solvents. *Chemosphere* **2018**, *201*, 361–369. [CrossRef]
19. Xu, X.; Li, L.; Yan, F.; Jia, Q.; Wang, Q.; Ma, P. Predicting solubility of fullerene C60 in diverse organic solvents using norm indexes. *J. Mol. Liq.* **2016**, *223*, 603–610. [CrossRef]
20. Danauskas, S.M.; Jurs, P.C. Prediction of C60 solubilities from solvent molecular structures. *J. Chem. Inf. Comput. Sci.* **2001**, *41*, 419–424. [CrossRef]
21. Toropov, A.A.; Rasulev, B.F.; Leszczynska, D.; Leszczynski, J. Additive SMILES based optimal descriptors: QSPR modeling of fullerene C60 solubility in organic solvents. *Chem. Phys. Lett.* **2007**, *444*, 209–214. [CrossRef]
22. Cheng, W.-D.; Cai, C.-Z. Accurate model to predict the solubility of fullerene C_{60} in organic solvents by using support vector regression. *Fuller. Nanotub. Carbon Nanostruct.* **2017**, *25*, 58–64. [CrossRef]
23. Ghasemi, J.B.; Salahinejad, M.; Rofouei, M.K. Alignment Independent 3D-QSAR Modeling of Fullerene (C_{60}) Solubility in Different Organic Solvents. *Fuller. Nanotub. Carbon Nanostruct.* **2013**, *21*, 367–380. [CrossRef]

24. Klamt, A. The COSMO and COSMO-RS solvation models. *Wiley Interdiscip. Rev. Comput. Mol. Sci.* **2011**, *1*, 699–709. [CrossRef]
25. Klamt, A. Conductor-like Screening Model for Real Solvents: A New Approach to the Quantitative Calculation of Solvation Phenomena. *J. Phys. Chem.* **1995**, *99*, 2224–2235. [CrossRef]
26. Klamt, A. *From Quantum Chemistry to Fluid Phase Thermodynamics and Drug Design*; Elsevier Science: Amsterdam, The Netherlands, 2005; ISBN 978-0-444-51994-8.
27. Klamt, A.; Eckert, F. COSMO-RS: A novel and efficient method for the a priori prediction of thermophysical data of liquids. *Fluid Phase Equilib.* **2000**, *172*, 43–72. [CrossRef]
28. Klamt, A. COSMO-RS for aqueous solvation and interfaces. *Fluid Phase Equilib.* **2016**, *407*, 152–158. [CrossRef]
29. Klamt, A.; Eckert, F.; Hornig, M.; Beck, M.E.; Brger, T.; Bürger, T. Prediction of aqueous solubility of drugs and pesticides with COSMO-RS. *J. Comput. Chem.* **2002**, *23*, 275–281. [CrossRef] [PubMed]
30. Loschen, C.; Klamt, A. New Developments in Prediction of Solid-State Solubility and Cocrystallization Using COSMO-RS Theory. In *Computational Pharmaceutical Solid State Chemistry*; John Wiley & Sons, Inc.: Hoboken, NJ, USA, 2016; pp. 211–233.
31. Eckert, F. Chapter 12. Prediction of Solubility with COSMO-RS. In *Developments and Applications in Solubility*; Royal Society of Chemistry: Cambridge, UK, 2007; pp. 188–200.
32. Tang, W.; Wang, Z.; Feng, Y.; Xie, C.; Wang, J.; Yang, C.; Gong, J. Experimental Determination and Computational Prediction of Androstenedione Solubility in Alcohol + Water Mixtures. *Ind. Eng. Chem. Res.* **2014**, *53*, 11538–11549. [CrossRef]
33. Schröder, B.; Freire, M.G.; Varanda, F.R.; Marrucho, I.M.; Santos, L.M.N.B.F.; Coutinho, J.A.P. Aqueous solubility, effects of salts on aqueous solubility, and partitioning behavior of hexafluorobenzene: Experimental results and COSMO-RS predictions. *Chemosphere* **2011**, *84*, 415–422. [CrossRef]
34. Schröder, B.; Santos, L.M.N.B.F.; Rocha, M.A.A.; Oliveira, M.B.; Marrucho, I.M.; Coutinho, J.A.P. Prediction of environmental parameters of polycyclic aromatic hydrocarbons with COSMO-RS. *Chemosphere* **2010**, *79*, 821–829. [CrossRef]
35. Schröder, B.; Santos, L.M.N.B.F.; Marrucho, I.M.; Coutinho, J.A.P. Prediction of aqueous solubilities of solid carboxylic acids with COSMO-RS. *Fluid Phase Equilib.* **2010**, *289*, 140–147. [CrossRef]
36. Schröder, B.; Martins, M.A.R.; Coutinho, J.A.P.; Pinho, S.P. Aqueous solubilities of five *N*-(diethylaminothiocarbonyl)benzimido derivatives at T = 298.15 K. *Chemosphere* **2016**, *160*, 45–53. [CrossRef] [PubMed]
37. Martins, M.A.R.; Silva, L.P.; Ferreira, O.; Schröder, B.; Coutinho, J.A.P.; Pinho, S.P. Terpenes solubility in water and their environmental distribution. *J. Mol. Liq.* **2017**, *241*, 996–1002. [CrossRef]
38. Mishra, D.S.; Yalkowsky, S.H. Ideal Solubility of a Solid Solute: Effect of Heat Capacity Assumptions. *Pharm. Res.* **1992**, *9*, 958–959. [CrossRef] [PubMed]
39. Yalkowsky, S.H. Solubility and Partitioning V: Dependence of Solubility on Melting Point. *J. Pharm. Sci.* **1981**, *70*, 971–973. [CrossRef] [PubMed]
40. Hildebrand, J.H. The temperature dependence of the solubility of solid nonelectrolytes. *J. Chem. Phys.* **1952**, *20*, 190–191. [CrossRef]
41. Neau, S.H.; Bhandarkar, S.V.; Hellmuth, E.W. Differential Molar Heat Capacities to Test Ideal Solubility Estimations. *Pharm. Res.* **1997**, *14*, 601–605. [CrossRef] [PubMed]
42. Nordström, F.L.; Rasmuson, Å.C. Prediction of solubility curves and melting properties of organic and pharmaceutical compounds. *Eur. J. Pharm. Sci.* **2009**, *36*, 330–344. [CrossRef] [PubMed]
43. Hildebrand, J.H. A Critique of the Theory of Solubility of Non-Electrolytes. *Chem. Rev.* **1949**, *44*, 37–45. [CrossRef] [PubMed]
44. Tosun, I. Solid-Liquid Equilibrium. In *The Thermodynamics of Phase and Reaction Equilibria*; Elsevier: Amsterdam, The Netherlands, 2013; pp. 509–549. ISBN 978-0-444-59497-6.
45. Moller, B. Activity of Complex Multifunctional Organic Compounds in Common Solvents. 2009. Available online: https://www.semanticscholar.org/paper/Activity-of-complex-multifunctional-organic-in-Moller/f48ee7d6b85a88e836bd1dadd344e238ca352d94 (accessed on 1 June 2019).
46. Mishra, D.S.; Yalkowsky, S.H. Solubility of organic compounds in non-aqueous systems: Polycyclic aromatic hydrocarbons in benzene. *Ind. Eng. Chem. Res.* **1990**, *29*, 2278–2283. [CrossRef]
47. Nordström, F.L.; Rasmuson, A.C. Determination of the activity of a molecular solute in saturated solution. *J. Chem. Thermodyn.* **2008**, *40*, 1684–1692. [CrossRef]

48. Svärd, M.; Rasmuson, A.C. (Solid + liquid) solubility of organic compounds in organic solvents—Correlation and extrapolation. *J. Chem. Thermodyn.* **2014**, *76*, 124–133. [CrossRef]

49. Marcus, Y.; Smith, A.L.; Korobov, M.V.; Mirakyan, A.L.; Avramenko, N.V.; Stukalin, E.B. Solubility of C_{60} Fullerene. *J. Phys. Chem. B* **2001**, *105*, 2499–2506. [CrossRef]

50. Hansen, C.M.; Smith, A.L. Using Hansen solubility parameters to correlate solubility of C_{60} fullerene in organic solvents and in polymers. *Carbon.* **2004**, *42*, 1591–1597. [CrossRef]

51. Jin, Y.; Cheng, J.; Varma-Nair, M.; Liang, G.; Fu, Y.; Wunderlich, B.; Xiang, X.D.; Mostovoy, R.; Zettl, A.K. Thermodynamic characterization of fullerene (C_{60}) by differential scanning calorimetry. *J. Phys. Chem.* **1992**, *96*, 5151–5156. [CrossRef]

52. Kamat, P.V.; Guldi, D.M.; Kadish, K.M. *Recent Advances in the Chemistry and Physics of Fullerenes and Related Materials: Proceedings of the Twelfth[sic] International Symposium*; Electrochemical Society: Pennington, NJ, USA, 1999; ISBN 1566772346.

53. Letcher, T.M.; Crosby, P.B.; Domanska, U.; Fowler, P.W.; Legon, A.C. Solubility of Buckminsterfullerene, C_{60}, in benzene and toluene. *S. Afr. J. Chem. Tydskr. Chem.* **1993**, *46*, 41–43.

54. Acree, W.; Chickos, J.S. Phase Transition Enthalpy Measurements of Organic and Organometallic Compounds and Ionic Liquids. Sublimation, Vaporization, and Fusion Enthalpies from 1880 to 2015. Part 2. C_{11}–C_{192}. *J. Phys. Chem. Ref. Data* **2017**, *46*, 013104. [CrossRef]

55. Kulkarni, P.P.; Jafvert, C.T. Solubility of C60 in solvent mixtures. *Environ. Sci. Technol.* **2008**, *42*, 845–851. [CrossRef]

56. Ruoff, R.S.; Malhotra, R.; Huestis, D.L.; Tse, D.S.; Lorents, D.C. Anomalous solubility behaviour of C_{60}. *Nature* **1993**, *362*, 140–141. [CrossRef]

57. Prausnitz, J.M.; Lichtenthaler, R.N.; de Azevedo, E.G. *Molecular Thermodynamics of Fluid-Phase Equilibria*; Prentice Hall PTR: Upper Saddle River, NJ, USA, 1999; ISBN 9780139777455.

58. Loos, P.-F.; Galland, N.; Jacquemin, D. Theoretical 0–0 Energies with Chemical Accuracy. *J. Phys. Chem. Lett.* **2018**, *9*, 4646–4651. [CrossRef]

59. Boetker, J.P.; Rantanen, J.; Arnfast, L.; Doreth, M.; Raijada, D.; Loebmann, K.; Madsen, C.; Khan, J.; Rades, T.; Müllertz, A.; et al. Anhydrate to hydrate solid-state transformations of carbamazepine and nitrofurantoin in biorelevant media studied in situ using time-resolved synchrotron X-ray diffraction. *Eur. J. Pharm. Biopharm.* **2016**, *100*, 119–127. [CrossRef] [PubMed]

60. Healy, A.M.; Worku, Z.A.; Kumar, D.; Madi, A.M. Pharmaceutical solvates, hydrates and amorphous forms: A special emphasis on cocrystals. *Adv. Drug Deliv. Rev.* **2017**, *117*, 25–46. [CrossRef] [PubMed]

61. Qu, H.; Kohonen, J.; Louhi-Kultanen, M.; Reinikainen, S.-P.; Kallas, J. Spectroscopic Monitoring of Carbamazepine Crystallization and Phase Transformation in Ethanol–Water Solution. *Ind. Eng. Chem. Res.* **2008**, *47*, 6991–6998. [CrossRef]

62. Harris, R.K.; Ghi, P.Y.; Puschmann, H.; Apperley, D.C.; Griesser, U.J.; Hammond, R.B.; Ma, C.; Roberts, K.J.; Pearce, G.J.; Yates, J.R.; et al. Structural Studies of the Polymorphs of Carbamazepine, Its Dihydrate, and Two Solvates. *Organic Process Res. Dev.* **2005**, *10*, 165. [CrossRef]

63. Li, Y.; Chow, P.S.; Tan, R.B.H.; Black, S.N. Effect of Water Activity on the Transformation between Hydrate and Anhydrate of Carbamazepine. *Organic Process Res. Dev.* **2008**, *12*, 264–270. [CrossRef]

64. Peerless, J.S.; Bowers, G.H.; Kwansa, A.L.; Yingling, Y.G. Fullerenes in Aromatic Solvents: Correlation between Solvation-Shell Structure, Solvate Formation, and Solubility. *J. Phys. Chem. B* **2015**, *119*, 15344–15352. [CrossRef] [PubMed]

65. Mitsari, E.; Romanini, M.; Qureshi, N.; Tamarit, J.L.; Barrio, M.; Macovez, R. C_{60} Solvate with (1,1,2)-Trichloroethane: Dynamic Statistical Disorder and Mixed Conformation. *J. Phys. Chem. C* **2016**, *120*, 12831–12839. [CrossRef]

66. Ye, J.; Barrio, M.; Céolin, R.; Qureshi, N.; Rietveld, I.B.; Tamarit, J.L. Van-der-Waals based solvates of C_{60} with CBr_2Cl_2 and $CBr_2(CH_3)_2$. *Chem. Phys.* **2016**, *477*, 39–45. [CrossRef]

67. Ye, J.; Barrio, M.; Negrier, P.; Qureshi, N.; Rietveld, I.B.; Céolin, R.; Tamarit, J.L. Orientational order in the stable buckminster fullerene solvate $C_{60} \cdot 2CBr_2H_2$. *Eur. Phys. J. Spec. Top.* **2017**, *226*, 857–867. [CrossRef]

68. Smith, A.L.; Walter, E.; Korobov, M.V.; Gurvich, O.L. Some Enthalpies of Solution of C_{60} and C_{70}. Thermodynamics of the Temperature Dependence of Fullerene Solubility. *J. Phys. Chem.* **2002**, *100*, 6775–6780. [CrossRef]

69. Sachsenhauser, T.; Rehfeldt, S.; Klamt, A.; Eckert, F.; Klein, H. Consideration of dimerization for property prediction with COSMO-RS-DARE. *Fluid Phase Equilib.* **2014**, *382*, 89–99. [CrossRef]
70. Cysewski, P. Prediction of ethenzamide solubility in organic solvents by explicit inclusions of dintermolecular interactions within the framework of COSMO-RS-DARE. *J. Mol. Liq.* **2019**, accepted.

Article

Riemann-Symmetric-Space-Based Models in Screening for Gene Transfer Polymers

Claudiu N. Lungu [1,*] and Ireneusz P. Grudzinski [2]

[1] Department of Chemistry, Faculty of Chemistry and Chemical Engineering, Babes-Bolyai University, 400028 Cluj, Romania
[2] Department of Applied Toxicology, Faculty of Pharmacy, Medical University of Warsaw, 02-097 Warsaw, Poland; ireneusz.grudzinski@wum.edu.pl
* Correspondence: lunguclaudiu5555@gmail.com

Received: 10 October 2019; Accepted: 27 November 2019; Published: 1 December 2019

Abstract: Today, gene transfer using polymers as transfer vectors is hardly studied. Some polymers have an excellent gene-carrying ability, but their cytotoxic and biocompatibility properties are not suitable for use. Thus, increased insight into the drug space of such structures is needed in the screening for suitable molecules. This study aimed to introduce a mathematical model of polymers suitable for genes transfer. In this regard, Riemann surfaces were used. The concerned polymers were taken from secondary published experimental data. The results show that symmetric Reimann spaces are suitable for further drug screening. The branch point values of Riemann surfaces are especially increased for the polymers suitable in gene transfer.

Keywords: polymers; gene transfer; Riemann space; chemical space; manifold; branching point

1. Introduction

The lack of gene-delivery vectors [1] is the main limiting factor in the field of gene personalized therapy [2]. Viral vectors have a low efficiency. Synthetic gene-delivery agents do not possess the required efficacy [3]. In recent years, a variety of capable polymers have been designed, specifically for gene delivery. Understanding the polymer gene-delivery mechanisms will help in the design of polymer-based gene-delivery systems, thus becoming essential tools for human gene therapy. The design criteria for the construction of useful delivery vectors are continuously evolving. Reverse drug design, fragment-based drug design, and virtual screening all failed in identifying a class candidate. Most of these methods explore the same drug space of a template compound. Thus, a mathematical model, expressed as a function, will enlarge the drug ability space. Such a model will go beyond the respective class of compounds. A way of generating such a model is to explore Cartesian coordinates, close contacts, dihedral angle values, and the internal coordinates of a certain molecule. Cartesian coordinates can provide a high starting point for projection in a vast range of mathematical spaces, compared with internal coordinates and close contacts. One can distinguish linear spaces and topological spaces. Linear spaces lack 3D dimensionality due to their algebraic nature. Linear operations performed in a linear space lead to straight lines. Their dimensionality is defined as the maximal number of independent vectors.

Topological spaces that are analytic in nature lead to continuous functions. A topological space is hard to define. An algebraic approach is used in most cases.

Polynomial equations and their geometric properties are used worldwide in algebraic geometry. A major characteristic that makes polynomial equations a valuable tool in topology is their definition from a basic arithmetic operation—addition and multiplication. This operation, when used, retrieves smooth and Riemannian manifolds.

Manifolds are a type of topological space that resembles Euclidian space. The concept of manifolds is a keystone in geometry, while it allows complex structures to be characterized from the perspective of the concept of local topological properties. The cartesian coordinates and dihedral angles of a structure can be viewed as carrying these topological properties.

A smooth manifold is not an actual space. Furthermore, topological manifolds are defined as smooth manifolds that are finite linear spaces. In other words, the surface of an ellipsoid can be represented as a smooth manifold.

Every real structure has its own Euclidean space representation. Riemann manifolds can also be considered to be Euclidean spaces.

Mathematical spaces, usually those in a complex analysis, coexist with drug spaces both in real roots and complex ones [4]. The significant backtracking of these techniques based on molecule topology is the actual force field, i.e., the level of theory used to optimize the respective compound.

Statistical and mathematical models are essential tools in predicting molecular properties. QSAR methodology uses a set of molecules in order to predict one distinct molecular property by computing a set of descriptors based on which a quantitative structure–activity relationship (QSAR) equation is generated. This equation is used to further predict specific properties. The molecular descriptors used in QSAR models are driven from experimental data like logP, molar refractivity, polarizability, and theoretical descriptors, which are generally symbolical representations of a molecule. The theoretical descriptors characterize the constitution of the molecule, structural fragments, the graph invariants, the 3D properties (molecule size, volume), and the properties derived from the grid base molecular force field (GRID), comparative molecular field analysis (CoMFA) and similar methods, respectively. Both types of descriptors, experimental and theoretical, are used in establishing a prediction model. Such models involve a specific chemical space.

Furthermore, the more diversified the molecular set is (i.e., the more the molecules differ in respect to their molecular formula and structure) the more the model will cover a vast chemical space. Thus, by using a diverse set of molecules one will obtain a better prediction model. In other words, the set of molecules used in this study are diverse (i.e., distinct molecular formula, distinct structure) so they can characterize accurately a chemical space, making the model applicable to novel molecules

In this study, an inverse methodology was proposed in order to identify polymers with gene transfer capabilities. Instead of building a QSAR model that will explore a fraction of the chemical space, the chemical space was investigated with the help of the Cartesian coordinates. The equations derived from the Cartesian coordinates were used to compute chemical spaces—Riemannian spaces.

Theoretically, a model built in such a way provides an extensive overview of the drug space, both in real numbers and in imaginary complex number spaces. Such equations (polynomial equations)/models (Riemann spaces) can work as "reverse engineering" and lead to suitable gene-carrying polymers, with good cytotoxicity and biocompatibility profiles.

2. Materials and Methods

A set of 29 monomers (Table 1), retrieved from the gene transfer literature [5], were used to compute a mathematical model. The in silico models in the mol2 format of polymer structures were generated using the Chemoffice 2005 software package [6]. The models were optimized at the 6-31G level of theory [7]. The cartesian coordinates (Figure 1) of polymer structures were used to develop derivate equations of coordinates (Tables 2 and 3), using Mathematica 5.0 software [8]. The Riemann surfaces for each derived coordinate equation were solved and computed using the same software (Mathematica). The branch point for each equation was computed and represented. A 2D (complex map) and a 3D (Riemann surface) were used to represent the results. In addition, a single hypothesis QSAR model was computed using the Schrodinger 2009 software package [9]. The pharmacophores were computed; having compounds 8 and 21 as templates, they were chosen because of their proven gene transfer properties. Moreover, a combined pharmacophore hypothesis was developed. The branching points were compared against the known gene transfer properties. The methodology is exemplified for compounds 8 and 21.

Table 1. Monomers with proven gene transfer capabilities.

Item	Monomers with Proven Gene Transfer Capabilities
1	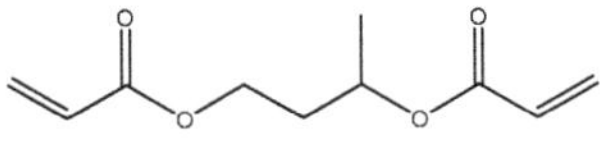
2	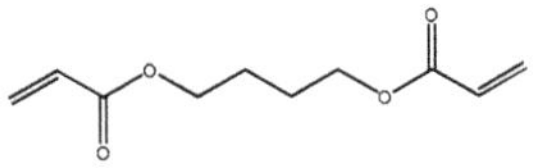
3	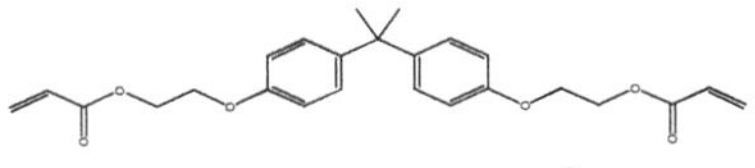
4	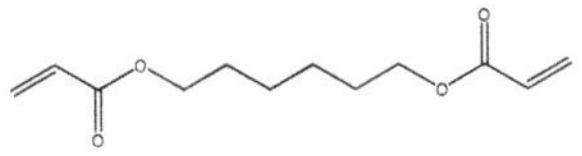
5	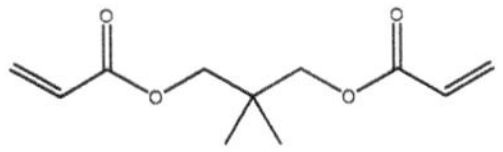
6	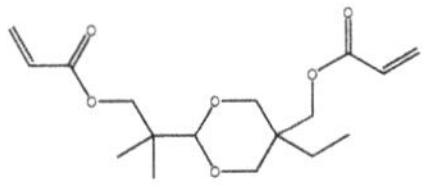
7	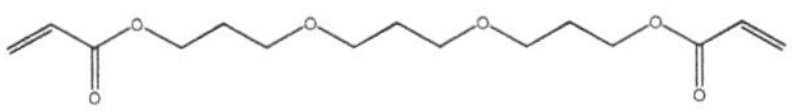
8	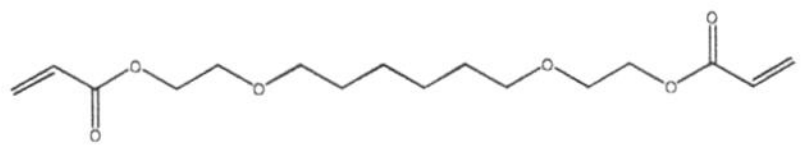
9	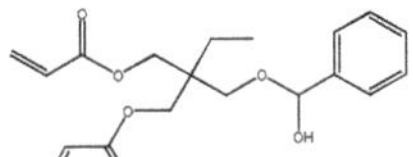
10	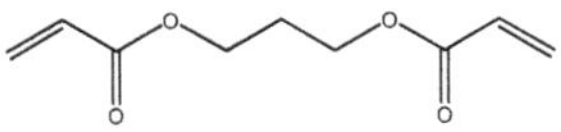
11	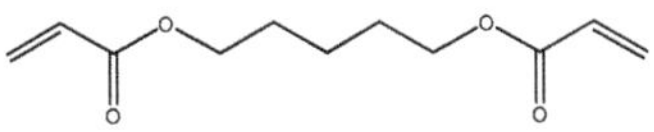
11	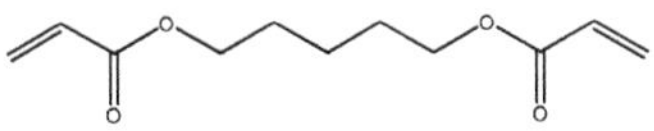
12	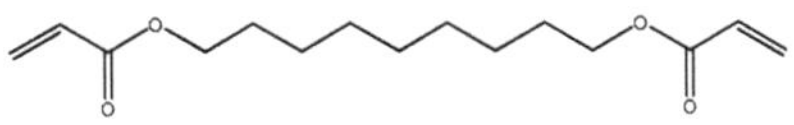
13	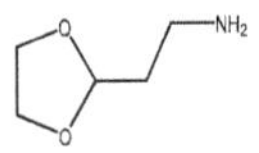
14	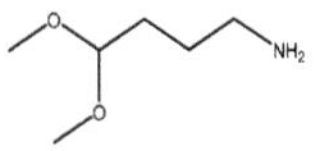
15	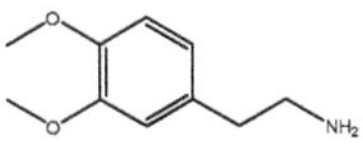

Table 1. *Cont.*

Item	Monomers with Proven Gene Transfer Capabilities
16	
17	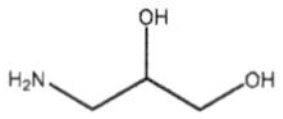
18	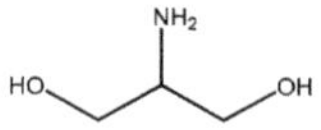
19	
20	
21	
22	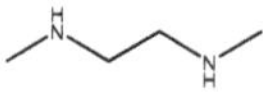
23	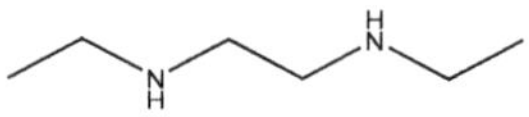
24	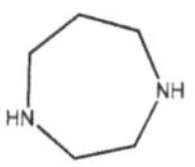
25	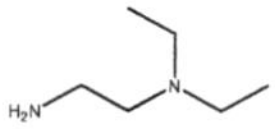
26	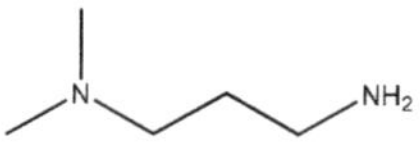
27	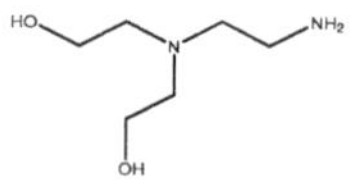
28	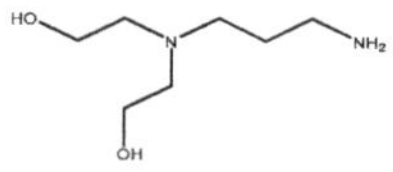
29	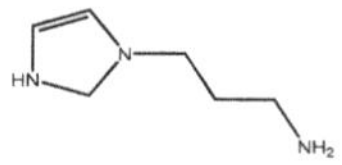

Table 2. Derived equations based on the Cartesian coordinates, where H is the independent variable.

	Monomer Equations
#	**Coordinate Based Equations**
1	$\partial_H(-13.4 + 19H)$
2	$\partial_H(-2.2 + 18.5H)$
3	$\partial_H(-13.0 + 28H)$
4	$\partial_H(-13.6 + 14H)$
5	$\partial_H(-2.4 + 15H)$
6	$\partial_H(-2.6 + 15H)$
7	$\partial_H(-4.9 + 17H)$
8	$\partial_H(-15.5 + 20H)$
9	$\partial_H(-4.5 + 23H)$
10	$\partial_H(-4.3 + 18H)$

Table 2. *Cont.*

	Monomer Equations	
#	**Coordinate Based Equations**
11	$\partial_H(-19.1 + 24H)$
12	$\partial_H(-3.4 + 19H)$
13	$\partial_H(-3.7 + 24H)$
14	$\partial_H(-2.4 + 21H)$
15	$\partial_H(-4.6 + 26H)$
16	$\partial_H(-4.0 + 2.8H)$
17	$\partial_H(6.8H)$
18	$\partial_H(-1.8 + 19H)$
19	$\partial_H(-1.9 + 795.2H)$
20	$\partial_H(-13.4 + 28H)$
21	$\partial_H(-28.7 + 28H)$
22	$\partial_H(-16.7 + 59H)$
23	$\partial_H(83.1H)$
24	$\partial_H(-10.8 + 31H)$
25	$\partial_H(25H)$
26	$\partial_H(31H)$
27	$\partial_H(-7.3 + 323.5H)$
28	$\partial_H(398.6H)$
29	$\partial_H(-5.8 + 49H)$

Table 3. Distinct equations for each compound based on the Cartesian coordinates.

#	**Equations for Computing Riemann Surfaces**
1	$z(-13.4 + 19z)\^(1/(z(-13.4 + 19z)))$
2	$z(-2.2 + 18.5z)\^(1/(z(-2.2 + 18.5z)))$
3	$z(-13.0 + 28z)\^(1/(z(-13.0 + 28z)))$
4	$z(-13.6 + 14z)\^(1/(z(-13.6 + 14z)))$
5	$z(-2.4 + 15z)\^(1/(z(-2.4 + 15z)))$
6	$z(-2.6 + 15z)\^(1/(z(-2.6 + 15z)))$
7	$z(-4.9 + 17z)\^(1/(z(-4.9 + 17z)))$
8	$z(-15.5 + 20z)\^(1/(z(-15.5 + 20z)))$
9	$z(-4.5 + 23z)\^(1/(z(-4.5 + 23z)))$
10	$z(-4.3 + 18z)\^(1/(z(-4.3 + 18z)))$
11	$z(-19.1 + 24z)\^(1/(z(-19.1 + 24z)))$
12	$z(-3.4 + 19z)\^(1/(z(-3.4 + 19z)))$
13	$z(-3.7 + 24z)\^(1/(z(-3.7 + 24z)))$
14	$z(-2.4 + 21z)\^(1/(-2.4 + 21z)))$
15	$z(-4.6 + 26z)\^(1/(z(-4.6 + 26z)))$
16	$z(-4.0 + 2.8z)\^(1/(z(-4.0 + 28z)))$
17	$z(6.8z)\^(1/(z(6.8z)))$
18	$z(-1.8 + 19z)\^(1/(z(-1.8 + 19z)))$
19	$z(-1.9 + 795.2z)\^(1/(z(-1.9 + 795.2z)))$
20	$z(-13.4 + 28z)\^(1/(z(-13.4 + 28z)))$
21	$z(-28.7 + 28z)\^(1/(z(-28.7 + 28z)))$
22	$z(-16.7 + 59z)\^(1/(z(-16.7 + 59z)))$
23	$z(83.1z)\^(1/(z(83.1z)))$
24	$z(-10.8 + 31z)\^(1/(z(-10.8 + 31z)))$
25	$z(25z)\^(1/(z(25z)))$
26	$z(31z)\^(1/(z(31z)))$
27	$z(-7.3 + 323.5z)\^(1/(z(-7.3 + 323.5z)))$
28	$z(398.6z)\^(1/(z(398.6z)))$
29	$z(-5.8 + 49z)\^(1/(z(-5.8 + 49z)))$

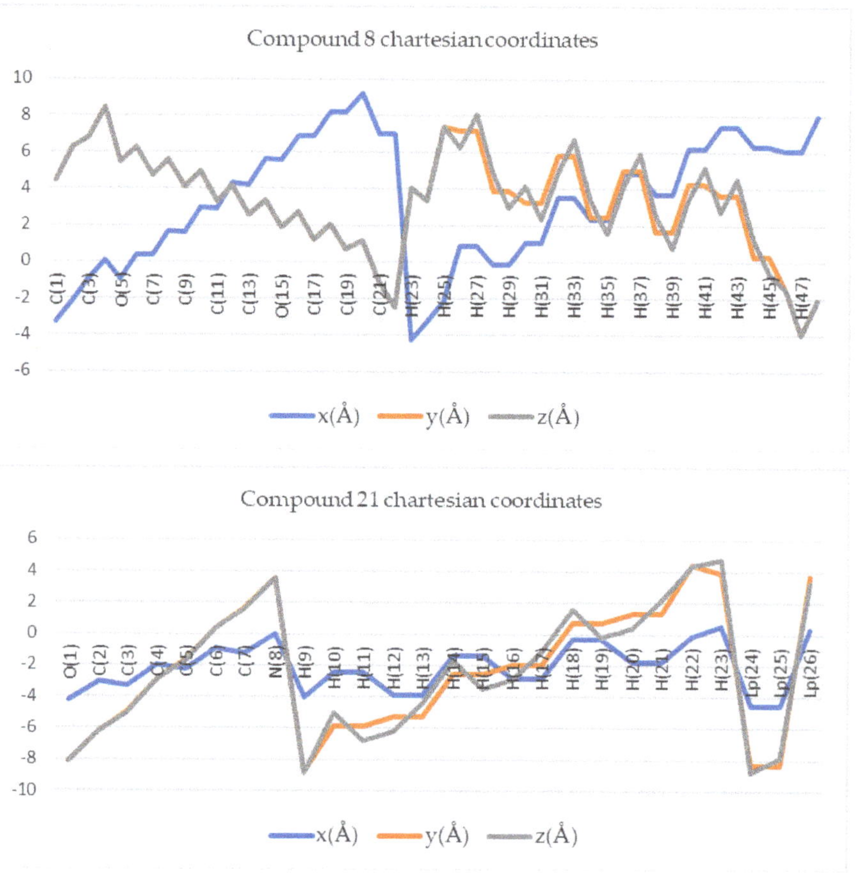

Figure 1. Cartesian coordinates of compounds 8 and 21.

3. Results

Table 2 lists the initial derivate equations generated using the Cartesian coordinates of each molecule.

The equations in Table 2 were transformed using the following formula:

$$\partial H(-a + bH) \rightarrow z(-a + bz)\hat{}(1/(z(-a + bz))) \tag{1}$$

where a = free term; b = coefficient; z = Lambert W function.

The equations for each polymer are shown in Table 3.

The Riemann surfaces obtained for compounds 8 and 21 that showed promising experimental results are shown in Figure 2.

The branching points computed for each structure are shown in Figure 3.

The QSAR models, developed as a single hypothesis for compounds 8 and 21, and the merge hypothesis are shown in Figure 4.

The Cartesian coordinates of each hypothesis are listed here. Hypothesis 1: D1D2H3H4; D1 (x −4.18, y 2.07, z −0.84); D2 (x −4.48, y −2.07, z 0.84); H3 (x −5.97, y 2.62, z 0.00); H4 (x −2.69 y −2.69, z 0.00).

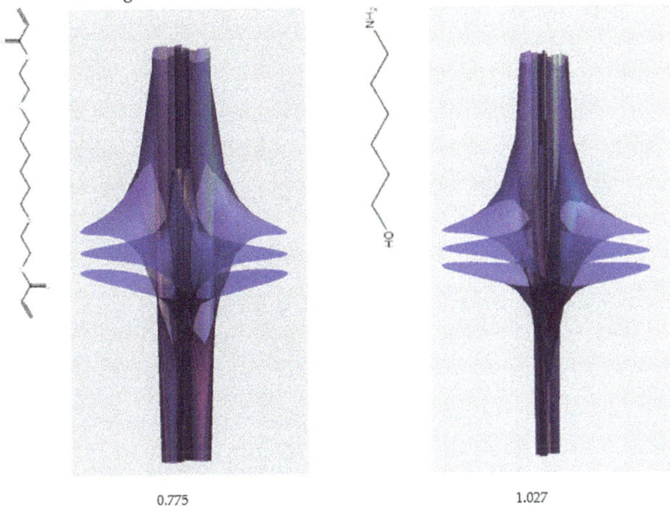

Figure 2. Riemann surfaces for compounds 8 and 21 respectively with their respective branching point values.

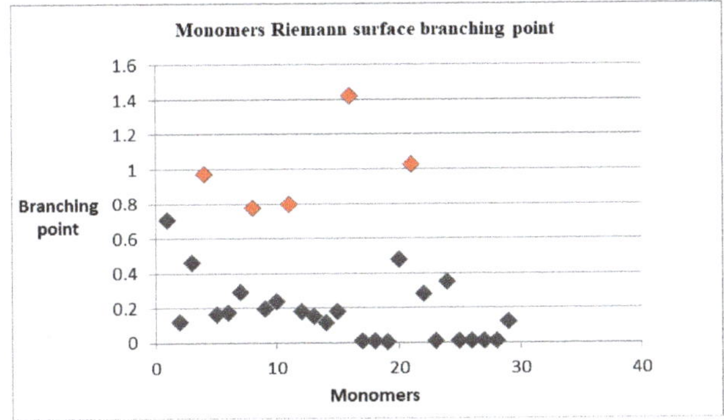

Figure 3. Branching points for the concerned monomers; points of structures with promising gene transfer profile, in red (see Supplementary Material S1).

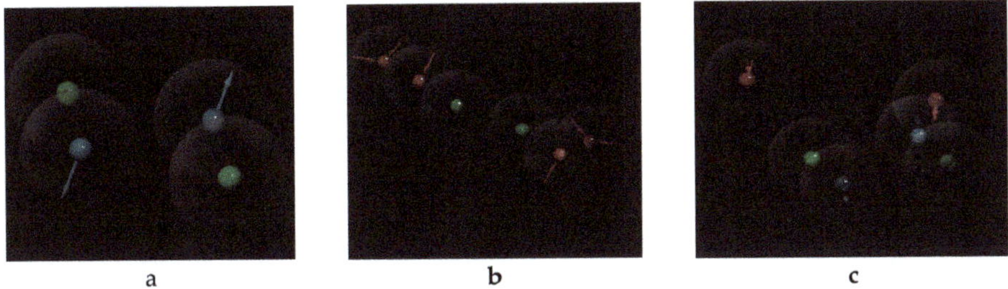

Figure 4. (**a**) Pharmacophore hypothesis based on molecule #8; (**b**) pharmacophore hypothesis based on molecule #21; (**c**) merge pharmacophore hypothesis (a + b): lyophilic (H-atom group, in green); hydrogen donor group (D, in blue); hydrogen acceptor group (A, in red).

Hypothesis 2: A1A2A3A4H5H6; A1 (x −3.11, y 3.66, z 0.00); A2 (x 2.06, y −3.25, z 0.00); A3 (x −2.32, y 5.77, z 0.00); A4 (x −2.32, y 5.77, z 0.00); H5 (−1.38, y 1.91, z 0.00); H6 (x 0.35, y −1.50, z 0.00).

Merge hypothesis: A1A2D2D3H5H6; A1 (x −3.11, y 3.66, z 0.00); A2 (x 2.06, y −3.25, z 0.00); D2 (x −4.48, y −2.07, z 0.34); D3 (x −4.18. y 2.07, z −0.84); D4 (x −4.48, y −2.07, z 0.84); H5 (x −5.97, y 2.62, z 0.00); H6 (x −2.69, y −2.62, z 0.00). As observed above, the merge hypothesis has common elements from both hypotheses 1 and 2, but also leaves out D1 and H4 from hypotheses 1 and A3 and A4 from hypothesis 2, respectively.

4. Discussion

Symmetric spaces are pseudo-Riemann manifolds [10]. A connected Riemann manifold is symmetric (space) if its curvature tensor is invariant. Broadly, a Riemann manifold is symmetric if each point exists as an isometry. Furthermore, every symmetric space is exhaustive.

Riemann symmetric spaces are considered in physics, mathematics, and chemistry. They have a central role in the homology theory. Examples of Riemann spaces include Euclidean spaces, hyperbolic spaces, and projective spaces. Riemann spaces are classified into the Euclidean type, the Compact type, and the Non-Compact type [11].

If the complex argument of a function can be mapped from a single point in the domain of multiple points, then the branching point of an analytical function is a point in the complex plane [12]. When z (branching point) = 0, under the power function $f(z) = za$, where "a" is a complex non-integer ("a" $\in$ C, with a $\nexists$ Z). Writing $z = ei\theta$ and taking θ in the interval 0 to 2π results in:

$$f(e^{oi}) = e^0 = 1; f(e^{2\pi i}) = e^{2\pi ia} \tag{2}$$

so that the values of the function $f(z)$; $z = (0;2\pi)$ are different [12–15].

In Riemann surfaces, the aspect of a branch point is defined for a holomorphic function when f: $X \rightarrow Y$, from a compact connected Riemann surface X to the compact Riemann surface Y (usually the Riemann sphere). If f is not constant, f will be a covering map onto its image at all but a finite number of points. The points of X, where the function f fails, are the bifurcations points of f, and the branch point is an image of a ramification point under f.

For any point P $\in$ X and Q = f(P) $\in$ Y, there are the holo-morphic local coordinates z for X near P and w for Y near Q, in terms of which the function $f(z)$ is given by $\omega = zk$, for some integer k. This integer is called the ramification index of P. Usually, the ramification index equals one; if the branching index is not equal to one, then P is, by definition, a ramification point, and Q is a branch point.

If Y is just the Riemann sphere, and Q is in the finite part of Y, then there is no demand to select particular coordinates. The ramification index (Equation (3)) can be determined explicitly from Cauchy's integral formula. Let γ be a simple rectifiable loop in X around P. The ramification index of f at P is (P-any point in the space with local holomorphic coordinates)

$$eP = \frac{1}{2\pi i} \int_y \frac{f\prime(z)}{f(z) - f(P)} dz \tag{3}$$

This integral is the number of times that f(y) winds over the point Q. As above, P is a ramification point, and Q is a branch point if eP > 1.

A Riemann surface is a surface-like composition that encloses the complex plane with infinitely many "sheets." These sheets can have very intricate structures and interconnections. Riemann surfaces are one way of representing multiple-valued functions; another way is represented by the branch cuts. The plot in Figure 1 shows the Riemann surfaces as the solutions of the equation:

$$[w(z)]\hat{}d + w(z) + z\hat{}(d-1) = 0 \tag{4}$$

with d = 2, 3, 4, and 5, where w(z) is the Lambert W-function.

The Riemann surface S of the function field K is the set of non-trivial discrete evaluations on K. Here, the set S corresponds to the ideals of the ring A of integers of K over z. Riemann surfaces provide a geometric visualization of the function elements and their analytical continuations.

Schwarz proved, at the end of the nineteenth century, that the automorphism group of a unified Riemann surface of genus g ≥ 2 is finite; then, Hurwitz showed that the group order is at most 84 (g − 1), where "g" is the genus.

In light of the computational results, polymers #4, 8, 11, 16 and 21, are the best candidates for feasible gene transfer. Experimentally, only polymers 8 and 21 showed good and acceptable results. A threshold regarding the branching point and its correlation with bioactivity was observed (Figure 2). A branching point of ~1 has an excellent correlation with gene transfer capability, cytotoxicity, and biocompatibility.

The QSAR single hypothesis models for compounds 8 and 21 revealed the importance of topology in performing bioactivity. The merge hypothesis presents a shared future for both compounds 8 and 21. Hydrogen atom accepting A-groups and donor D-groups are critical in the chemical space of compounds [16]. Furthermore, the pharmacophores demonstrated a relatively diverse set of functional groups for each hypothesis, findings that suggest a lack of specificity in describing a common pharmacophore.

Lastly, using the characteristics of symmetric mathematical space [17], one can characterize distinct molecular properties, as shown in Table 4, where the Riemman spaces for compounds 8 and 21 are represented with their complex aspects (see Supplementary Material S2) [18–20].

Table 4. Riemann surfaces: 1—function plots, 2—complex space plots, 3—real part 3D plot, 4—imaginary part 3D plot, 5—branching number, and 6—polymers (#).

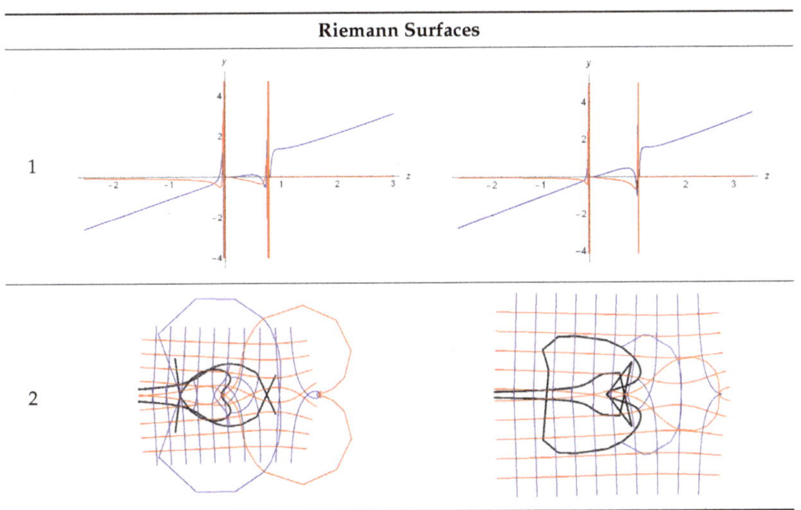

Table 4. *Cont.*

	Riemann Surfaces	
3		
4		
5	0.7754	1.027
6	# 8	# 21

5. Conclusions

A suitable mathematical model, based on Riemann surfaces can be built in order to screen and characterize polymers with gene transfer properties. The branching point of Riemann surfaces can aid in assessing such surfaces, their connection with the drug space and their connective properties.

Supplementary Materials: The following are available online at http://www.mdpi.com/2073-8994/11/12/1466/s1. S2—Riemann surfaces.

Author Contributions: C.N.L. established the conceptual framework, produced the results and discussion, and assembled the paper. I.P.G. performed the literature screening of concerned compounds and methods and contributed the references and the result discussion.

Funding: This work was supported by the GEMNS project granted in the European Union's Seventh Framework Program under ERA-NET EuroNanoMed II (European Innovative Research and Technological Development Projects in Nanomedicine).

Acknowledgments: We are especially indebted to all reviewers for graciously going through these drafts and making numerous substantive corrections and suggestions about how to improve the quality of the paper. We are especially indebted to Sohan Jheeta (of The Network of Researchers on the Chemical Evolution of Life, NoR CEL, www.nor-cel.com) for graciously going through several drafts and making numerous substantive suggestions about how to improve the quality of the paper.

Conflicts of Interest: The authors declare no conflict of interest.

Symmetry **2019**, *11*, 1466

References

1. Mary, B.; Maurya, S.; Kumar, M.; Bammidi, S.; Kumar, V.; Jayandharan, G.R. Molecular engineering of Adeno-associated virus capsid improves its therapeutic gene transfer in murine models of hemophilia and retinal degeneration. *Mol. Pharm.* **2019**, *16*, 4738–4750. [CrossRef] [PubMed]
2. Aartsma-Rus, A.; van Putten, M. The use of genetically humanized animal models for personalized medicine approaches. *Dis. Model. Mech.* **2019**, *13*, 041673. [CrossRef] [PubMed]
3. Contin, M.; Garcia, C.; Dobrecky, C.; Lucangioli, S.; D'Accorso, N. Advances in drug delivery, gene delivery and therapeutic agents based on dendritic materials. *Future Med. Chem.* **2019**, *11*, 1791–1810. [CrossRef] [PubMed]
4. Richard, B.; Ho-Fung, C.; Jóhannes, R. Characteristics of known drug space. Natural products, their derivatives and synthetic drugs. *Eur. J. Med. Chem.* **2010**, *45*, 5646–5652.
5. Anderson, D.G.; Peng, W.; Akinc, A.; Naushad, H.; Kohn, A.; Pandera, R.; Langer, R.; Sawicli, A.J. A polymer library approach to suicide gene therapy for cancer. *Proc. Natl. Acad. Sci. USA* **2004**, *9*, 16028–16033. [CrossRef] [PubMed]
6. Beata, S.; Mircea, V.D.; Mihai, V.P.; Ireneusz, P.G. Docking linear ligands to glucose oxidase. *Int. J. Mol. Sci.* **2016**, *17*, 1796.
7. Nguyen, L.H.; Nguyen, T.H.; Truong, T.N. Quantum Mechanical-Based Quantitative Structure-Property Relationships for Electronic Properties of Two Large Classes of Organic Semiconductor Materials: Polycyclic Aromatic Hydrocarbons and Thienoacenes. *ACS Omega* **2019**, *4*, 7516–7523. [CrossRef] [PubMed]
8. *Mathematica*, Version 8.0; Wolfram Research, Inc.: Champaign, IL, USA, 2010.
9. *Schrödinger Release Prime*; Schrödinger LLC: New York, NY, USA, 2009.
10. Akhiezer, D.N.; Vinberg, E.B. Weakly symmetric spaces and spherical varieties. *Transf. Groups* **1999**, *4*, 3–24. [CrossRef]
11. Wolf Joseph, A. *Spaces of Constant Curvature*, 5th ed.; McGraw–Hill: New York, NY, USA, 1999.
12. Conway John, B. *Functions of One Complex Variable*, 2nd ed.; Springer: Berlin/Heidelberg, Germany, 1978.
13. Arfken, G. *Mathematical Methods for Physicists*, 3rd ed.; Academic Press: Orlando, FL, USA, 1985; pp. 397–399.
14. Morse, P.M.; Feshbach, H. *Methods of Theoretical Physics, Part I*; McGraw-Hill: New York, NY, USA, 1953; pp. 391–392, 399–401.
15. Trott, M. *The Mathematica Guidebook for Programming*; Springer: Berlin/Heidelberg, Germany, 2004; pp. 188–191.
16. Weisstein, E.W. "Branch Point" from MathWorld—A WolframWeb Resource. Available online: http://mathworld.wolfram.com/BranchPoint.html (accessed on 5 December 2018).
17. Majumdar, S.; Basak, S.C.; Lungu, C.N.; Diudea, M.V.; Grunwald, G.D. Mathematical structural descriptors and mutagenicity assessment: A study with congeneric and diverse datasets. *SAR QSAR Environ. Res.* **2018**, *29*, 579–590. [CrossRef] [PubMed]
18. Petitjean, M. A definition of symmetry. *Symmetry Cult. Sci.* **2007**, *18*, 99–119.
19. Lungu, C.N. C-C chemokine receptor type 3 inhibitors: Bioactivity prediction using local vertex invariants based on thermal conductivity layer matrix. *Stud. Univ. Babes-Bolyai Chem.* **2018**, *63*, 177–188. [CrossRef]
20. Lungu, C.N.; Diudea, M.V.; Putz, M.V. Ligand shaping in induced fit docking of MraY inhibitors. Polynomial discriminant and Laplacian operator as biological activity descriptors. *Int. J. Mol. Sci.* **2018**, *18*, 1377. [CrossRef] [PubMed]

MDPI
St. Alban-Anlage 66
4052 Basel
Switzerland
Tel. +41 61 683 77 34
Fax +41 61 302 89 18
www.mdpi.com

Symmetry Editorial Office
E-mail: symmetry@mdpi.com
www.mdpi.com/journal/symmetry